● 清华大学土木工程系列教材

土木工程学术论文写作与报告

赵志宏 编著

清华大学出版社
北京

内容简介

本书系统地阐述了土木工程学术论文写作与报告应遵循的基本原则；重点介绍了论文撰写、投稿、返修等写作环节与海报制作、口头演讲等交流环节的细节技巧；并通过大量示例提供了操作指南。

全书分为方法篇、实战篇两部分。方法篇包括论文写作、图表制作、写作范式、论文投稿、学术会议、其他写作、学术伦理、文献检索与管理等；实战篇包括公式编辑指南、图片制作指南、表格编辑指南、常用句式与词汇、投稿流程简介、土木工程领域期刊缩写、论文检索与管理软件使用指南、学术共同体行为准则等。

本书可作为高等学校土木工程、水利工程、海洋工程等专业的研究生教材，也可作为其他工科专业或工程技术人员的写作参考书。

版权所有，侵权必究。举报：010-62782989，beiqinquan@tup.tsinghua.edu.cn。

图书在版编目（CIP）数据

土木工程学术论文：写作与报告 / 赵志宏编著. —北京：清华大学出版社，2021.1（2022.8重印）
清华大学土木工程系列教材
ISBN 978-7-302-56674-8

Ⅰ. ①土… Ⅱ. ①赵… Ⅲ. ①土木工程－英语－论文－写作－研究生－教材 Ⅳ. ①TU

中国版本图书馆 CIP 数据核字(2020)第 203638 号

责任编辑：秦　娜　王　华
封面设计：陈国熙
责任校对：王淑云
责任印制：丛怀宇

出版发行：清华大学出版社
网　　址：http://www.tup.com.cn，http://www.wqbook.com
地　　址：北京清华大学学研大厦 A 座　　邮　编：100084
社 总 机：010-83470000　　邮　购：010-62786544
投稿与读者服务：010-62776969，c-service@tup.tsinghua.edu.cn
质量反馈：010-62772015，zhiliang@tup.tsinghua.edu.cn

印 装 者：天津鑫丰华印务有限公司
经　　销：全国新华书店
开　　本：185mm×260mm　　印　张：12.75　　字　数：309 千字
版　　次：2021 年 1 月第 1 版　　印　次：2022 年 8 月第 2 次印刷
定　　价：39.80 元

产品编号：079187-01

前　言

2018年5月17日，在首场"清华名师教学讲坛"上，清华大学校长邱勇宣布：将在2018级新生中开设"写作与沟通"必修课程，计划到2020年，该课程将覆盖所有本科生，并力争面向研究生提供课程和指导。总体来看，写作和沟通能力不足，不仅是本科生培养中有待加强之处，也是很多硕士、博士研究生的短板。国外著名大学［如斯坦福大学（Stanford University）、加州大学伯克利分校（UC Berkeley）等］通常都为研究生开设类似的课程，名称一般为Writing in the Sciences或Scientific Writing，而且这些课程在国外大学都是研究生的必修课。作者在瑞典皇家工学院（Kungliga Tekniska Högskolan，KTH）攻读博士学位期间也曾学习了必修课程"Writing Scientific Papers"。近年来，学术英语已成为各高校研究生教育的新热点，比如清华大学各院系开设的关于科技论文写作的研究生课程就有20多门。关于科技论文写作的相关教材也很多，既有专门从事语言教学的教师编写的通用英语写作教材，也有从事一线科研工作的学者编写的具有专业特色的论文写作与交流指导用书。但是，还没有专门针对土木工程专业研究生学术写作与交流的教材。

学术写作与交流能力已成为土木工程专业研究生培养中的关键一环。清华大学土木工程系于2014年秋开设"科技论文写作与交流"研究生公共课，并得到清华大学研究生教改项目"土水学院'科技论文写作与交流'课程建设"的支持，本书基于过去6年的教学实践编写而成。本书不讨论如何做好的研究，关注的是如何写出好的论文，故本书所介绍的内容都是基于读者已经做出了"漂亮"的研究成果。

本书的主要特点：①写作与沟通并重，主要内容不仅涉及论文写作、投稿、修改等各个环节，还包括参加学术会议时幻灯片与海报的制作、演讲方法等交流环节；②既包含学术论文写作与交流的一般性原则与技巧，还通过大量示例提供了操作指南；③本书主要面向土木工程专业的研究生和高年级本科生，书中示例多是作者自己或作者的同事、学生亲身经历的，具有较强的针对性。需要说明的是，本书阐述的学术论文写作与交流的一般原则也适用于其他工程学科。

清华大学图书馆的战玉华老师自2018年起为"科技论文写作与交流"课程讲授"文献检索与管理利用""文献管理软件使用方法"等内容。本书的第8章和第15章由战老师独立撰写完成，作者由衷地感谢战老师在"科技论文写作与交流"课程建设和本书编写出版过程中给予的大力支持。

在本书的编写过程中，作者参考了国内外众多科技论文写作与交流方面的教材和著作，比如 *How to write and publish a scientific paper*（第7版）（Robert A. Day和Barbara Gastel）、*Academic writing for graduate students*（第3版）（John M. Swales和Christine B. Feak）、《新学者融入世界科坛》（何毓琦）等，这些著作在确定本书的结构、内容、风格等方面使作者受益良多，受篇幅所限，不能一一提及。

本书的出版得到了清华大学研究生院、清华大学土木水利学院的支持与资助。作者课题组的博士后和研究生刘桂宏、陈跃都、尚德磊、彭欢、郭铁成、窦子豪、徐浩然、陈思聪、林涛、王佳铖等参与了第9～16章的撰写与图表绘制。作者还要感谢过去6年选修"科技论文写作与交流"课程的208名研究生，他们通过教学评估、当面交流等形式为课程也为本书提出了很多建议。

在本书的编写过程中，作者力求在内容上兼顾土木工程及其他相近学科，所选案例贴近低年级研究生和高年级本科生的写作需求，但由于作者水平有限，书中不当与错误之处在所难免，敬请广大读者及专家批评指正。

作者

2020年5月13日

于清华园

目　　录

绪论 ·· 1

 0.1 正确的"论文观" ·· 2
 0.2 良好的写作习惯 ··· 2
 0.3 高效的交流能力 ··· 3

方　法　篇

第 1 章　论文写作 ·· 7

 1.1 学术论文类型 ·· 7
 1.2 论文主要组成部分 ··· 7
 1.3 标题撰写要点 ·· 8
 1.4 作者信息撰写要点 ··· 9
 1.5 亮点撰写要点 ··· 11
 1.6 摘要撰写要点 ··· 11
 1.7 关键词撰写要点 ··· 12
 1.8 引言撰写要点 ··· 13
 1.9 方法撰写要点 ··· 14
 1.10 结果撰写要点 ··· 16
 1.11 讨论撰写要点 ··· 17
 1.12 结论撰写要点 ··· 18
 1.13 致谢撰写要点 ··· 18
 1.14 附录撰写要点 ··· 19
 1.15 参考文献撰写要点 ·· 19
 1.16 补充材料撰写要点 ·· 21
 1.17 论文写作流程 ··· 21

第 2 章　图表制作 ··· 22

 2.1 表格 vs 图片 ·· 22
 2.2 表格制作要点 ··· 25
 2.3 图片 ·· 25
 2.4 图形摘要 ··· 29

第 3 章 写作范式 ·········· 31
3.1 学术论文风格 ·········· 31
3.2 语态 ·········· 32
3.3 时态 ·········· 33
3.4 用词 ·········· 35
3.5 常见错误 ·········· 36

第 4 章 论文投稿 ·········· 39
4.1 期刊选择 ·········· 39
4.2 论文投稿 ·········· 40
4.3 审稿 ·········· 42
4.4 论文修改 ·········· 46
4.5 论文发表 ·········· 47

第 5 章 学术会议 ·········· 48
5.1 会议论文 ·········· 48
5.2 口头报告 ·········· 50
5.3 学术海报 ·········· 51
5.4 交流礼仪 ·········· 52

第 6 章 其他写作 ·········· 53
6.1 综述论文 ·········· 53
6.2 学术短评 ·········· 54
6.3 学术简历与求职信 ·········· 56
6.4 项目申请书 ·········· 58
6.5 推荐信 ·········· 59

第 7 章 学术伦理 ·········· 61
7.1 论文版权 ·········· 61
7.2 发表伦理 ·········· 63
7.3 案例分析 ·········· 65

第 8 章 文献检索与管理 ·········· 67
8.1 文献检索基础知识 ·········· 67
8.2 常用文献资源 ·········· 67
8.3 文献检索 ·········· 71
8.4 文献管理与利用 ·········· 75
8.5 小结 ·········· 76

实 战 篇

第 9 章 公式编辑指南 79
    9.1 公式编辑注意事项 79
    9.2 公式的自动编号和引用 80

第 10 章 图片制作指南 83
    10.1 数据图 83
    10.2 流程图 120
    10.3 示意图 123
    10.4 图像 125
    10.5 地图 132

第 11 章 表格编辑指南 134
    11.1 软件介绍 134
    11.2 操作步骤 135

第 12 章 常用句式与词汇 139
    12.1 引言 139
    12.2 方法 140
    12.3 结果 141
    12.4 讨论 142
    12.5 词汇 142

第 13 章 投稿流程简介 144
    13.1 前期准备（Preparation） 144
    13.2 稿件投递（Submission） 146

第 14 章 土木工程领域期刊缩写 153

第 15 章 论文检索与管理软件使用指南 160
    15.1 课题介绍 160
    15.2 论文检索步骤 160
    15.3 论文检索结果及分析 164
    15.4 文献管理 168
    15.5 小结 172

第 16 章 学术共同体行为准则 173
    16.1 出版商（以下为爱思维尔的出版商申明） 173

  16.2 主编 ··· 174
  16.3 审稿人 ··· 175
  16.4 作者 ··· 176
附录 A 中英文对照表 ··· 179
附录 B CURRICULUM VITAE ·· 182
参考文献 ·· 190

绪　　论

随着我国科教事业的蓬勃发展，我国的学者在国际学术界的影响力与日俱增，发表高水平的学术论文和报告、参与学术团体的组织和管理已成为衡量研究人员学术成就与影响力的重要指标之一。持续发表高水平论文不仅是体现学者学术贡献的主要途径之一，也是获得研究资助和职称晋升的重要前提（图0.1）。对于青年学者或研究生来说，在论文撰写、投稿、参加学术会议、加入国际学术组织等过程中，须熟悉并遵守国际惯例，以更高效地宣传自己的学术成果、更快地融入"学术圈"，避免因文化差异导致的误会。

图 0.1　持续发表高水平论文贯穿研究人员学术生涯（WWW.PHDCOMICS.COM）

The evolution of intellectual freedom—学术"自由"之路；I'm going to research whatever I want! —我要研究我想要研究的！I'm going to research whatever my professor wants! —我要研究导师想要研究的！I'm going to research whatever my tenure committee wants! —我要研究长聘委员会想要研究的！I'm going to research whatever my grant committee wants! —我要研究基金评审委员会想要研究的！I'm going to research whatever I—我要研究（我想要研究的）；Research In Peace—学术"自由"；Before grad school—本科生；Grad student—研究生；Assistant professor—助理教授；Tenured professor—长聘教授；Emeritus professor—退休教授

土木工程（civil engineering）一级学科涵盖结构工程、岩土工程、市政工程、暖通空调、防灾减灾、桥梁隧道等多个二级学科，其显著特点是实践性和综合性都很强。相对于其他学科，土木工程诞生早，与人类生存关系密切。随着时代的发展和科技的进步，土木工程学科的研究正在从定性向定量、经验向智能、传统向信息等维度转变，而且与材料、机械、能源等学科的交叉也越来越广泛。在此背景下，我们认为非常有必要编写一本针对土木工程专业研究生学术写作与交流的教材，帮助青年学者和研究生树立正确的论文观、培养良好的写作习惯与交流能力，以促进土木工程的发展。

0.1　正确的"论文观"

高水平论文：一本期刊被科学引文索引(science citation index,SCI)收录并不保证该期刊有较高的学术影响力,在 SCI 收录的期刊上发表一篇论文并不完全反映该论文的学术水平,因此,仅以被 SCI 收录的论文数目作为评论依据既不客观也不全面。目前比较公认的是一篇论文被其他 SCI 论文引用的次数,它的确在某种程度上反映了论文的质量。但是不同学科领域,论文引用次数存在很大差别,学科间横向比较意义通常不大。

学术不端零容忍：切勿因为项目结题、毕业、职称晋升等压力去发表毫无独创性的论文。不重复发表或将原本完整的"故事"(研究成果)拆分为若干篇论文,以增加论文数目。切忌编造、篡改或选择性地发表试验数据。

0.2　良好的写作习惯

研究中：边做研究边读文献,跟踪学术最新动向;边读文献边记录,建立常用句式、单词素材库。

撰写中：理清论文结构,与合作者仔细讨论、反复修改(图 0.2),并与参考文献中典型论文进行对比改进。

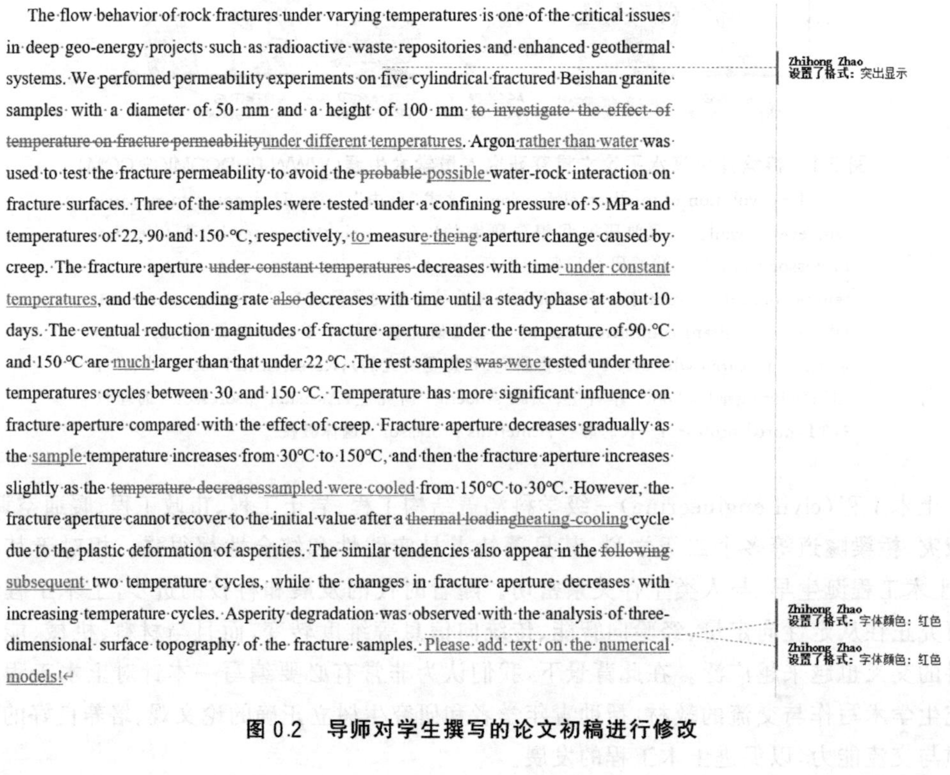
The flow behavior of rock fractures under varying temperatures is one of the critical issues in deep geo-energy projects such as radioactive waste repositories and enhanced geothermal systems. We performed permeability experiments on five cylindrical fractured Beishan granite samples with a diameter of 50 mm and a height of 100 mm ~~to investigate the effect of temperature on fracture permeability~~under different temperatures. Argon rather than water was used to test the fracture permeability to avoid the ~~probable~~ possible water-rock interaction on fracture surfaces. Three of the samples were tested under a confining pressure of 5 MPa and temperatures of 22, 90 and 150 °C, respectively, to measure the~~ing~~ aperture change caused by creep. The fracture aperture ~~under constant temperatures~~ decreases with time under constant temperatures, and the descending rate ~~also~~ decreases with time until a steady phase at about 10 days. The eventual reduction magnitudes of fracture aperture under the temperature of 90 °C and 150 °C are much larger than that under 22 °C. The rest samples ~~was~~ were tested under three temperatures cycles between 30 and 150 °C. Temperature has more significant influence on fracture aperture compared with the effect of creep. Fracture aperture decreases gradually as the sample temperature increases from 30°C to 150°C, and then the fracture aperture increases slightly as the ~~temperature decreases~~sampled were cooled from 150°C to 30°C. However, the fracture aperture cannot recover to the initial value after a ~~thermal loading~~heating-cooling cycle due to the plastic deformation of asperities. The similar tendencies also appear in the ~~following~~ subsequent two temperature cycles, while the changes in fracture aperture decreases with increasing temperature cycles. Asperity degradation was observed with the analysis of three-dimensional surface topography of the fracture samples. Please add text on the numerical models!

Zhihong Zhao
设置了格式：突出显示

Zhihong Zhao
设置了格式：字体颜色：红色

Zhihong Zhao
设置了格式：字体颜色：红色

图 0.2　导师对学生撰写的论文初稿进行修改

投稿前：论文完成后可搁置几天，然后再通读一遍查找形式、拼写错误，确认无误后再投稿。

投稿后：与编辑保持良好的沟通，正确面对评审意见，利用修改机会再次提高论文质量。

0.3 高效的交流能力

早准备：尽早准备好会议交流的 PPT，请有经验的合作者帮忙修改提高。

勤练习：将每张 PPT 所要讲述的内容写下来、背下来，避免做报告时因临时"找词"而紧张。

考虑到土木工程研究生参与学术交流活动的一般顺序，并兼顾水利工程、采矿工程等专业研究生的需求，本书内容分为方法篇与实战篇两大部分。方法篇注重写作与交流的基本原则，共包含 8 章：第 1 章阐述常规科技论文的组成部分以及各部分的写作要点和注意事项；第 2 章给出高质量图、表的要求与制作方法；第 3 章讨论英文写作的范式与细节问题；第 4 章详细介绍论文投稿过程中可能遇到的各种问题及处置方法；第 5 章给出参加学术会议的流程与做高质量学术报告的注意事项；第 6 章概述其他的英文写作；第 7 章强调学术论文写作与交流中应注意的伦理规范等；第 8 章简要介绍科技文献的检索与管理方法。实战篇注重写作与交流的具体操作指南，共包含 8 章：第 9 章介绍公式编辑方法；第 10 章介绍各类图片的制作方法及软件使用指南；第 11 章介绍各类表格的制作方法及软件使用指南；第 12 章提供科技论文写作中常用的句式与词汇；第 13 章介绍投稿流程；第 14 章罗列土木工程领域主要的英文期刊及其缩写；第 15 章介绍论文检索与管理软件操作技巧；第 16 章介绍学术共同体行为准则。

书后附录罗列本书中主要名词的中英文对照表以及个人简历范例。

方 法 篇

武松打虎

第 1 章 论文写作

学术论文的核心组成部分通常包括引言(Introduction)、方法(Methodology)、结果(Results)、讨论/结论(Discussion/Conclusion)四部分,即 IMRD 结构。除此之外,还包括标题(Title)、作者信息(Author information)、亮点(Highlights)、摘要(Abstract)、关键词(Keywords)、图(Figures)、表(Tables)、致谢(Acknowledgement)、参考文献(References)、附录(Appendices)、补充材料(Supplementary materials)等其他要素。本章重点介绍如何撰写学术论文的文字部分,包括各部分的作用、撰写原则、注意事项等,并针对性地提供示例供读者参考。

1.1 学术论文类型

学术论文是描述原始研究成果并经同行评审后出版的书面载体。常见的论文类型主要有:

(1) 论文:学术论文的最常见形式,英文称作 Research article、Original paper、Full paper 等,其长度在 5000~10 000 个单词。

(2) 短文:简明扼要地阐述新理论、新技术、新发现等,英文称作 Technical note 或 Short communication,其长度在 3000 个单词以内。

(3) 案例研究:描述新理论或新技术在典型工程中的应用效果,英文称作 Case study,其长度在 5000~10 000 个单词。

(4) 讨论:针对新近发表的某一学术论文进行简要且公开的讨论、回复和勘误/更正等,一般要求在该论文发表 3 个月内提交编辑部,英文称作 Letter、Closures、Errata,其长度在 1000 个单词以内。

(5) 综述:就某一时期某研究领域的研究进展进行总结、分析和评价,英义称作 Review article,在第 6 章中会对其单独介绍。

(6) 会议论文:为出席学术会议而撰写的论文,通常收入论文集,英文称作 Conference paper,在第 5 章中会对其单独介绍。

1.2 论文主要组成部分

按照学术期刊的通常要求,论文的组成部分可分为必备和可选两类(图 1.1)。必备部分是指所有学术论文共同的必需要素,而可选部分则是根据期刊制定或作者需要额外添加的要素。

(1) 标题：明确论文的研究主题。

(2) 作者信息：给出全部作者的姓名、单位、联系方式等信息。

(3) 亮点(可选)：概述 3~5 条论文的主要学术贡献，即创新点。

(4) 摘要：高度概括论文各章节的主要内容，即浓缩版的论文。

(5) 关键词：帮助索引器和搜索引擎找到相关论文的工具。

(6) 正文：包含引言、方法、结果、讨论/结论等核心章节。

(7) 致谢(可选)：对研究工作提供支持的单位和个人表达感谢。

(8) 附录(可选)：提供与研究方法、研究结果相关的详细资料。

(9) 参考文献：提供与研究工作紧密相关的文献，有助于读者理解相较于前人工作的创新点、研究结果的可靠性、研究主题与研究领域的相关性等。

(10) 补充材料(可选)：提供主要研究结果之外的原始研究数据或全部原始数据等。

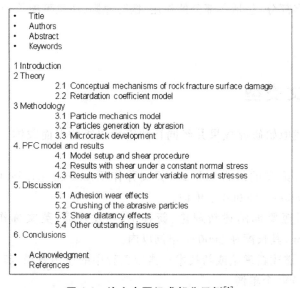

图 1.1　论文主要组成部分示例[1]

1.3　标题撰写要点

标题是学术论文阅读量最大的组成部分。读者首先通过论文标题了解论文主题，如感兴趣才会阅读摘要和整篇论文。一篇标题不准确的论文可能永远不会被其真正的目标读者阅读。撰写论文标题的最主要原则是**用尽可能少的词语来准确全面地描述论文主题**。具体包括以下要点：

(1) 标题长度以 10~20 个单词为宜。

例：Numerical modeling of stability of fractured reservoir bank slopes subjected to water-rock interactions[2]

例：Evidence for tensile faulting deduced from full waveform moment tensor inversion during the stimulation of the Basel enhanced geothermal system[3]

（2）标题须包含论文的研究对象、研究工况、研究方法（**室内试验、数值模拟、理论推导或是现场实测为主**）等重要信息，但标题通常不是一个完整的句子。

　　　　　　　　研究方法　　　　　　　　研究工况　　　　　　　研究对象
例：**Experimental investigation** on **wetting-induced weakening** of **sandstone joints**[4]

（3）标题不宜包含缩写词、化学式、专有名称、名称和术语等，但当标题过长且不影响同行理解的情况下也可酌情使用上述缩略语。

例：DEM modeling of interaction between the propagating fracture and multiple pre-existing cemented discontinuities in shale[5]

DEM 为 Discrete Element Method 的缩写，为岩土领域学者所熟知，故此处采用该缩写以使标题简洁紧凑。

（4）悬挂标题可用于强调论文的研究方法或研究角度。

例：Impact of stress on solute transport in a fracture network：A comparison study[6]

例：Direct and distributed strain measurements inside a shotcrete lining：Concept and realization[7]

（5）系列标题常用于若干紧密相关的学术论文，用于强调每篇论文的研究重点。

例：Sorbing tracer experiments in a crystalline rock fracture at Äspö（Sweden）：1. Experimental results and micro-scale characterization of retention properties[8]

Sorbing tracer experiments in a crystalline rock fracture at Äspö（Sweden）：2. Transport model and effective parameter estimation[9]

Sorbing tracer experiments in a crystalline rock fracture at Äspö（Sweden）：3. Effect of microscale heterogeneity [10]

1.4　作者信息撰写要点

论文作者及其排序明确了责任和信誉。学术论文最易撰写的部分就是作者署名，即按对论文的贡献如实列出作者姓名、工作单位、联系方式等信息（图1.2）。但是，论文署名也是容易出现学术不端/不当的地方，此方面的内容将在第 7 章介绍，本节主要介绍如何真实准确地提供作者信息。

（1）论文作者应且仅应包括为研究的想法、方案设计、执行和结果分析等做出贡献的研究人员。

（2）作者署名顺序通常遵循对研究成果的贡献大小排序。

例：学者 A：完成试验方案的详细设计。

技术员 B：严格按照 A 设计的试验方案完成试验。

如果试验成功，撰写学术论文时 A 为唯一作者，可在致谢中对 B 予以感谢。

如果试验不成功，B 根据经验提供一些建议，A 同意后，重新试验成功，撰写学术论文时

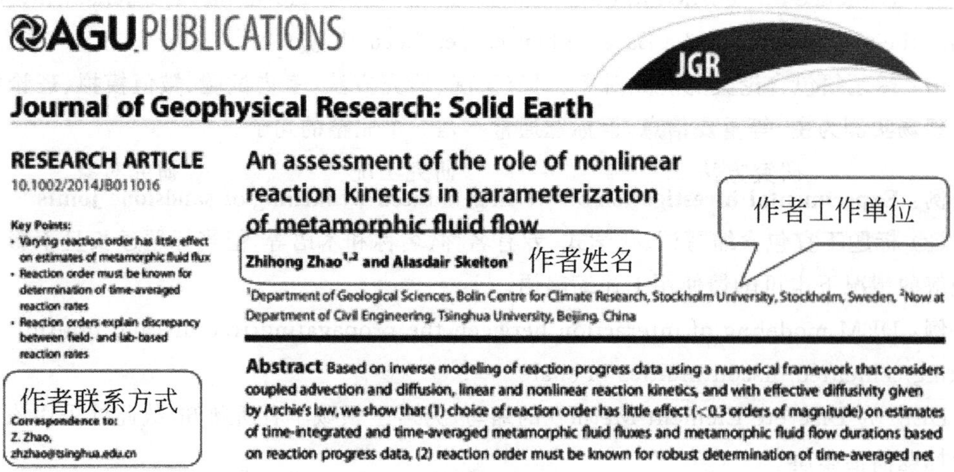

图 1.2 作者信息示例[11]

A 和 B 都应为作者。

如果试验不成功，B 根据经验提供一些建议，A 同意后，重新试验成功。A 和 B 共同撰写论文草稿，学者 C 审阅论文草稿并给出了一些建设性的意见，在这种情况下，A 和 B 为论文作者，可在致谢中对 C 予以感谢。如果 C 的贡献足够显著，并征得 C 同意后，也可考虑将 C 列为合作者之一，比如 A、B、C 或 A、C、B。

（3）部分期刊要求明确说明论文作者的学术贡献，确保每位合作者确应署名，而且有助于读者根据需要联系哪位作者获取哪类信息。

例：A.S. conceived the project design; H.K., A.S., and H.R.G. provided expertise on water-rock interaction in Iceland; H.H. collected the samples; M.A., C.M.M., and H.S. performed the chemical analysis; G.S. and M.A. performed the PHREEQCI modeling; S.J. and E.S. provided expertise on seismicity; I.K. performed the statistical analysis; A.S. wrote the paper with input from all co-authors.[12]

例：Alessandro F. Rotta Loria: Conceptualization, Methodology, Software, Validation, Investigation, Supervision, Writing - review & editing. José V. Catalá Oltra: Data curation, Visualization, Investigation. Lyesse Laloui: Supervision, Writing-review & editing.[13]

（4）作者姓名格式：通常遵循名在前、姓在后的格式，推荐在学术论文中按全名格式署名。

例：赵志宏署名为 Zhihong Zhao。注意避免在不同论文中分别使用 Zhihong Zhao、Zhi-Hong Zhao、Z. Zhao 等不同格式。

（5）作者工作单位应包括单位详细信息。

例：Department of Civil Engineering, Tsinghua University, Beijing, China

（6）通信作者：负责论文投稿、修改、与主编沟通、回复读者询问等。通信作者应提供电子邮箱（E-mail），并保证在出版过程中以及出版后都能联系到（图 1.2）。

1.5 亮点撰写要点

亮点应准确描述最重要的研究发现以及所使用的新方法（如果有），不应试图涵盖论文所有的概念、研究结果或结论。准确简练的亮点有助于提高论文的可发现性。

（1）亮点一般为 3~5 条，且多数期刊对亮点有明确的长度限制（图 1.3），比如爱思维尔（Elsvier）的期刊要求亮点不超过 85 个字符（含空格）。

（2）亮点既可以是完整的句子，也可以是像"标题"一样的短语。

（3）亮点应采用一般现在时。

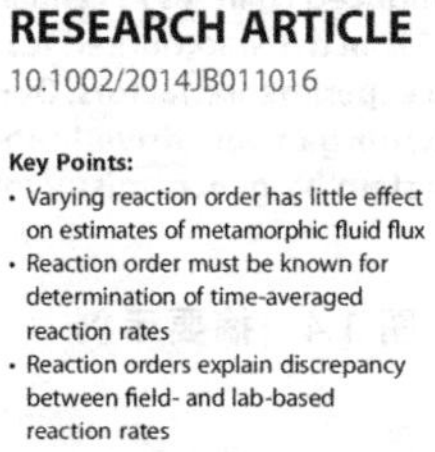

RESEARCH ARTICLE
10.1002/2014JB011016

Key Points:
- Varying reaction order has little effect on estimates of metamorphic fluid flux
- Reaction order must be known for determination of time-averaged reaction rates
- Reaction orders explain discrepancy between field- and lab-based reaction rates

(a) [11]

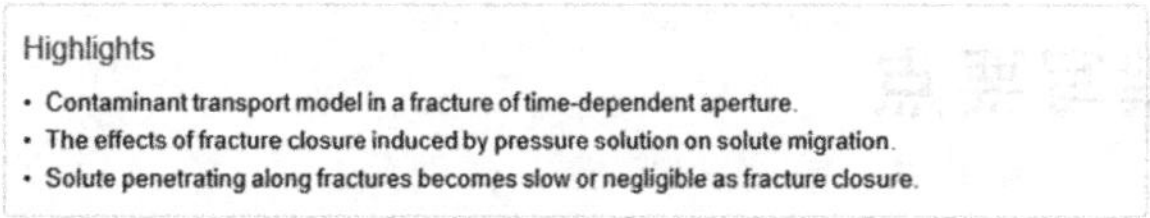

Highlights
- Contaminant transport model in a fracture of time-dependent aperture.
- The effects of fracture closure induced by pressure solution on solute migration.
- Solute penetrating along fractures becomes slow or negligible as fracture closure.

(b) [14]

图 1.3 亮点示例

1.6 摘要撰写要点

摘要应对学术论文各主要部分进行高度概括。高度凝练且信息全面的摘要有助于读者快速准确地了解论文的创新点，确定是否有必要阅读全文。

（1）摘要应简要介绍研究背景和意义，清晰界定研究内容，准确介绍研究方法，客观呈现主要研究结果与结论。摘要不应涉及任何正文中没有陈述的研究结果或结论。摘要一般不引用参考文献、不添加图表。

（2）摘要通常为单独一个段落，且不应超过期刊规定的长度（一般为 200~250 字）。

（3）摘要中描述已完成的工作时应采用一般过去时，其他内容可采用一般现在时或现在完成时（图 1.4）。

在图 1.4 所示示例中，第 1 句至第 3 句：研究背景（一般现在时或现在完成时）；第 4 句至第 6 句：研究方法和主要结果（一般过去时）；第 7 句：主要结论（一般现在时）；第 8 句：研究意义（一般现在时）。

SCIENCE ADVANCES | RESEARCH ARTICLE

ECOLOGY

A keystone microbial enzyme for nitrogen control of soil carbon storage

Ji Chen[1,2,3], Yiqi Luo[4,5]*, Kees Jan van Groenigen[6], Bruce A. Hungate[5], Junji Cao[2,7], Xuhui Zhou[8,9], Rui-wu Wang[1]

Agricultural and industrial activities have increased atmospheric nitrogen (N) deposition to ecosystems worldwide. N deposition can stimulate plant growth and soil carbon (C) input, enhancing soil C storage. Changes in microbial decomposition could also influence soil C storage, yet this influence has been difficult to discern, partly because of the variable effects of added N on the microbial enzymes involved. We show, using meta-analysis, that added N reduced the activity of lignin-modifying enzymes (LMEs), and that this N-induced enzyme suppression was associated with increases in soil C. In contrast, N-induced changes in cellulase activity were unrelated to changes in soil C. Moreover, the effects of added soil N on LME activity accounted for more of the variation in responses of soil C than a wide range of other environmental and experimental factors. Our results suggest that, through responses of a single enzyme system to added N, soil microorganisms drive long-term changes in soil C accumulation. Incorporating this microbial influence on ecosystem biogeochemistry into Earth system models could improve predictions of ecosystem C dynamics.

图 1.4 摘要示例[15]

1.7 关键词撰写要点

关键词应包括论文的研究对象、研究方法和研究条件等关键信息。每条关键词通常由 1~4 个单词组成，且关键词数目一般不超过 5 个（图 1.5）。

ARTICLE INFO

Article history:
Received 20 September 2012
Received in revised form
27 August 2013
Accepted 24 December 2013

Keywords:
Solute transport
Pressure solution
Rock fracture
Time-dependent aperture

(a)[14]

ARTICLE INFO

Article history:
Received 29 October 2016
Received in revised form
2 May 2017
Accepted 8 May 2017
Available online 13 May 2017

Keywords:
Geothermal energy extraction
Coupled thermo-hydraulic
Discrete fracture
Finite element method
Parameters analyses

(b)[16]

图 1.5 关键词示例

1.8 引言撰写要点

引言应提供充分的研究背景与研究现状,并在此基础上明确点出研究不足(research gap),使读者无须参考其他相关文献即可了解论文拟解决的科学或技术问题。根据需要,也可给出所选研究方法的理由以及研究结果与结论。一般应在引言中定义要使用的专业术语或缩写。

Fractures in rock provide the dominant conductive pathways for fluids to move through the Earth's crust, influencing a wide range of subsurface human activities including the extraction of hydrocarbons, the sequestration of green house gases and the protection of aquifers. However, fractures are intrinsically heterogeneous and easily modified by natural and human processes. They may be altered geochemically and deformed under stress, affecting fluid flow rates that can vary across many orders of magnitude. 〉 研究背景

For hydrologic purposes, a fracture may be viewed as a quasi-two-dimensional (2D) network of void spaces through which fluids flow. Volumetric flow rates are controlled by the size and spatial distribution of the apertures of the void space[1-3]. On the other hand, the mechanical properties of a fracture are controlled by the asperities, which are the discrete points of contact between the two fracture surfaces. Apertures and asperities are complementary aspects of the same fracture geometry and each represents an influence network. The connected apertures define a percolation network, whereas the discrete points of contact are connected through the rock matrix as a separate stress network that controls the mechanical deformation of a fracture. These two networks combine to provide a full description of the fracture geometry, which is the nexus between the hydraulic and mechanical properties of a fracture. However, as demonstrated by many investigations, fracture geometry is complex and sensitive to alterations[4-7].

Recently, Petrovitch et al.[8] demonstrated numerically that fracture-specific stiffness is an effective parameter that captures the deformed topology of a fracture and can be used as the basis of a scaling relationship for fluid flow through fractures. This was an important development, because fracture-specific stiffness can be estimated from seismic wave attenuation and velocity[9-14]. The Petrovitch relationship holds for random distributions of weakly correlated apertures, but it remained an open question whether this relationship would be universal, that is, whether it would apply to spatially correlated aperture distributions or to fractures with preferential erosion that causes channelized flow. 〉 研究不足

In this study we demonstrate that a scaling relationship exists that accounts for spatial correlations in fracture aperture distributions and also captures the behaviour of channelized flow within a fracture. This relationship can be used as a guide for incorporating the appropriate behaviour of fractures into continuum models of the subsurface where stresses vary with depth. This relationship also provides a path forward to the ultimate goal of remotely monitoring fluid flow or relative fluid flow among fractures in the Earth's subsurface. 〉 研究目标及意义

图 1.6 摘要示例[17]

(1) 引言通常包含3个自然段以上，分别阐述研究背景、研究现状、研究目标，其中研究现状可以细化为多个自然段(图1.6)。

(2) 引言主要描述已有的研究成果，故多采用一般现在时态。

(3) 如引言涉及大量的文献总结时，可以考虑用表格的形式呈现(图1.7)。

例：

Table 1 Summarization of laboratory results about thermal influences on mechanical properties of various rocks

References	Rock type	Test temperature	Sample size	Elastic modulus	Compressive strength	Tensile strength
Duclos and Paquet (1991)	Basalt	20–900			Increase-decrease	
Araújo et al. (1997)	Sandstone	20–145	5 cm (φ) × 10 cm	Increase	Decrease	
Xu and Liu (2000)	Granite	20–600	2 cm (φ) × 4 cm	Increase-decrease	Increase-decrease	
Madland et al. (2002)	Chalk (wet)	20–130	3.7 cm (φ) × 7.4 cm	Decrease	Decrease	Decrease
	Chalk (Dry)			Increase	Increase	Increase
Liang et al. (2006)	Salt	20–240	ISRM SM	Decrease	Increase	
Rao et al. (2007)	Sandstone	20–300	5 cm (φ) × 5 cm 5 × 5 × 5 cm	Increase-decrease	Increase-decrease	Increase-decrease
Dwivedi et al. (2008)	Granite	30–160	ISRM SM	Decrease-increase	Decrease-increase	Decrease
Zhang et al. (2009)	Marble	20–800	2 cm (φ) × 4.5 cm	Decrease generally	Decrease generally	
	Limestone			Decrease generally	Decrease generally	
	Sandstone			Decrease generally	Fluctuating	
Vishal et al. (2011)	Khondalite	30–250	5.4 cm (φ) × 3.4 cm			Increase-decrease
Luo and Wang (2011)	Mudstone	20–750	2 × 2×5 cm³	Increase	Increase	
Zhao et al. (2012)	Coal	20–500	20 cm (φ) × 40 cm			
	Granite		20 cm (φ) × 40 cm	Decrease		
Ranjith et al. (2012)	Sandstone	25–950	2.3 cm (φ) × 4.6 cm	Increase-decrease	Increase-decrease	
Sriapai et al. (2012)	Salt	4–182	5.4 × 5.4 × 5.4 cm³	Decrease	Decrease	Decrease
Zhang et al. (2014)	Mudstone	25–800	2 cm (φ) × 4.5 cm	Increase-decrease	Increase-decrease	
Homand-Etienne and Houpert (1989)[a]	Granite	20–600	5 cm (φ) × 10 cm	Decrease	Decrease	Decrease
Ferrero and Marini (2001)[a]	Marble	0–600	ASTM SM	Decrease	Not affected	
Du et al. (2004)[a]	Granite	20–800	5 cm (φ) × 10 cm	Decrease generally	Decrease generally	
Lion et al. (2005)[a]	Limestone	20–250	3.7 cm (φ) × 7.4 cm	Decrease	Increase	
Wu et al. (2005)[a]	Sandstone	20–1200	5 cm (φ) × 10 cm	Decrease generally	Decrease generally	
Qiu and Lin (2006)[a]	Granite	20–800	5 cm (φ) × 10 cm	Decrease	Decrease	
Zhu et al. (2006)[a]	Tuff	20–800	5 cm (φ) × 10 cm	Decrease	Decrease	
	Granite			Decrease	Decrease	
	Breccia			Decrease generally	Decrease	
Koca et al. (2006)[a]	Marble	20–700	ISRM SM	Decrease	Decrease	
Xu et al. (2009)[a]	Granite	25–1300	2.5 cm (φ) × 5 cm	Decrease	Decrease	
Keshavarz et al. (2010)[a]	Gabbro	25–1000	4 cm (φ) × 9 cm	Decrease	Decrease	
Chen et al. (2012)[a]	Granites	20–1000	4 cm (φ) × 8 cm	Decrease	Decrease	
Brotóns et al. (2013)[a]	Calcarenite	60–105	5 cm (φ) × 12.5 cm	Decrease	Decrease	

[a] Indicates that the rock samples experienced thermal treatment (temperature increased from room temperature to the peak values, maintained for some time and decreased to room temperature), otherwise the tests were carried out for samples without cooling.

图1.7 引言中采用表格总结大量文献[18]

1.9 方法撰写要点

虽在引言中已提及所采用的研究方法，但在方法这一节中应给出关于研究方法的细节介绍，要使读者根据所描述的研究方法就可以复现论文的研究结果。如果引言中没有陈述研究方法的选择依据，那么在本节开头还应简要介绍可采用的研究方法，以及选择某一种或某几种研究方法的理由。

(1) 如涉及试验材料，应该准确给出材料技术规格、数量、来源和制备方法(图1.8)。

(2) 如果是新的研究方法(未发表过的)，则必须提供该研究方法的所有细节；如果该方法之前已发表，作者可提供必要的原始参考文献，不必赘述(图1.9)。

(3) 研究方法主要描述作者已完成的工作，故此节中大部分内容应该用一般过去时(图1.10)。

2.1. Sample Preparation

The material used for the present study is a commercially available microfine calcium sulphate cement mortar named flowstone (KPM Industries, Canada). The main compositions of flowstone are bassanite ($2CaSO_4 \cdot H_2O$) and calcite ($CaCO_3$), and it is an appropriate replica material for the study of rock friction and shear-induced asperity degradation (Tatone, 2014; supporting information Table S1). After curing, flowstone is mechanically similar to homogeneous limestones such as the Indiana limestone (Fjar et al., 2011).

Two flowstone samples (i.e., FL-I and FL-II) were prepared and tested in this study (see supporting information for detailed sample preparation steps). Each flowstone sample was prepared to be cylindrical with 32 mm in length and 12 mm in diameter. Then the samples were transversely split into two semisamples by means of 3-point bending test (American Society for Testing and Materials, 2002). This created matching rough surfaces between the semisamples that were used as the synthetic fault (Figures 1a and 1b). Then a hole 2.5 mm in diameter and 3 mm in depth was drilled along the sample axis on the center of each top semisample. The results on FL-I suggested that at unconfined condition, fractures penetrated into the sample body and split the sample. In order to examine the influence of this secondary fracturing, FL-II was further prepared before testing by applying retaining 1 mm thick aluminum jackets 2 mm above and below the fault (Figure 1b).

图 1.8　材料制备示例[19]

2.3. Experiment Setup and Procedure

The experiments were setup and performed following the steps described by Zhao et al. (2017; Figure 1e). For FL-I, a initial normal load of 280 N was applied, which resulted in a nominal normal stress (σ_n) of approximately 2.5 MPa. During the test, the top semisample was rotated incrementally with three 3° steps (Rot. I and II) and four 6° steps (Rot. III to VI). Experiment conducted on FL-II was carried out in a similar way but with initial σ_n = 1.8 MPa and twice the rotation step sizes. During each rotation, the top semisample was accelerated to the velocity of 3°/s in 0.1 s and stopped almost instantaneously when the desired amount of rotation was reached. The resultant averaged slip velocities for the FL-I and FL-II tests were 0.19 and 0.16 mm/s, respectively; and the resultant averaged acceleration for the FL-I and FL-II tests were 1.9 and 1.6 mm/s², respectively.

Normal force (N) and torque (M) were recorded at a sampling rate of 250 kHz, and the rotation distance was recorded every 0.06° of rotation. Note that the torque measurement signal saturated at full-scale (i.e., 1.5 N·m) during Rot. II of the FL-I test, and to avoid this from reoccurring, the voltage range of the torque acquisition channel was doubled for the successive tests.

The rotation step size was arbitrarily chosen, and after each incremental rotation step, we acquired a μCT data set. The small rotation step size allowed for in situ and in operando imaging of the gradual morphological evolution of the sample.

图 1.9　试验装置与流程[19]

Data analysis

We evaluated the effects of N additions by the natural log of the response ratio (ln R), a metric commonly used in meta-analysis (20, 39, 40).

非新方法，提供参考文献

$$\ln R = \ln\left(\frac{X_N}{X_C}\right) = \ln(X_N) - \ln(X_C) \qquad (1)$$

with X_C and X_N as the arithmetic mean values of the variables in the ambient and N addition treatments, respectively. The variances (v) of ln R are calculated by

$$v = \frac{S_N^2}{n_N X_N^2} + \frac{S_C^2}{n_C X_C^2} \qquad (2)$$

with n_C and n_N as the replicate numbers and S_C and S_N as the SDs for ambient and N addition treatments, respectively.

Meta-analysis was conducted using the "rma.mv" function in the R package "metafor" (http://cran.r-project.org/web/packages/metafor/index.html). Because several papers contributed more than one response ratio, we included the variable "publication" as a random factor (39, 40). The effects of N addition were considered significant if the 95% confidence interval did not overlap with zero. The results were reported as percentage change with N addition [that is, 100 × ($e^{\ln R}$ – 1)] to ease interpretation.

The meta-analytic models were selected by using the same approach as in van Groenigen et al. (39) and Terrer et al. (40). Briefly, we analyzed all possible combinations of the studied factors in a mixed-effects meta-regression model using the "glmulti" package in R (www.metafor-project.org/doku.php/tips:model_selection_with_glmulti). The importance of each predictor was expressed as the sum of Akaike weights for models that included this factor, which can be considered as the overall support for each variable across all models. A cutoff of 0.8 was set to differentiate between essential and nonessential predictors. We evaluated the impacts of soil pH, soil C/N, soil texture (clay content), and N-induced changes in plant productivity, soil pH, soil C/N, and microbial community on soil C storage using linear regression analysis in R.

图 1.10　研究方法示例[15]

1.10　结果撰写要点

研究结果应描述具有代表性或规律性的数据，不应重复研究方法中描述过的试验细节，也不应重复展示获得的数据结果。这个章节大多数采用过去时。

(1) 撰写研究结果应遵循看"图"/"表"说话的方式，客观准确地描述获得的研究结果，而不是简单重复"图"/"表"显而易见的信息。

(2) 研究结果主要描述已获得的数据，故此节中大部分内容应该用一般过去时（图1.11）。

(3) 结果部分最显著的特点是文字应与"图"/"表"一一对应，相辅相成（图1.12）。如果

3.4. Results

The best fits of equations (9a) and (9b) to reaction progress data with $n = 1.0$, 1.6, 2.0, and 2.7 are shown for four of the 14 sills that were studied in Figure 1. These were chosen as representative of the wide span of N_{Pe}–N_D combinations of the studied sills (i.e., high N_{Pe}–high N_D, low N_{Pe}–high N_D, high N_{Pe}–low N_D, and low N_{Pe}–low N_D). The parameters obtained from these best fits are presented in Table 1. For each sill, best fit concentration curves for all four preset reaction orders were similar. This is reflected by near-identical values of R^2 that expresses goodness of fit (Table 1). In other words, goodness of fit was unaffected by nonlinearity of the reaction rate law. In Figure 2, the best fit parameters obtained under $n = 1.0$ and $n = 2.7$ are compared. Best fit values of $v\phi t$ were unaffected by choice of reaction order, and the average value of $v\phi t$ for the 14 sills was 20 m³ m⁻² for linear reaction kinetics and 23 m³ m⁻² for nonlinear ($n = 2.7$) reaction kinetics. Similarly, calculated values of $v\phi$ and t were unaffected by choice of reaction order (Figure 2 and Table 1). For $n = 1.0$, k_{eff} was $10^{-9.76}$ s⁻¹, whereas for $n = 2.7$, k_{eff} was $10^{-7.84}$ s⁻¹. This result is further emphasized in Figure 3 where best fit estimates of $v\phi t$, $v\phi$, t, and k_{eff} are plotted against reaction order (n) and in Table 2 which shows errors arising by assuming linear reaction kinetics if $n = 2.7$.

图 1.11　研究结果示例 1[11]

RESULTS

Averaged across all studies, N addition significantly increased soil C storage by 11.0%. N addition significantly increased cellulase activity by 15.2% and repressed LME activity by 12.8% (Fig. 1A). Changes in soil C storage with N addition were negatively correlated with N suppression of LME activity, such that N-induced suppression of LME activity was associated with increases in soil C content (Fig. 1B). This negative relationship held over a range of ecosystems and N addition methods (figs. S2 and S3), although it was not significant for studies with high soil C/N ratios (>21.4; fig. S4). The response of LME activity explained 40.4% of the variation in soil C storage to N addition. In contrast, the effects of N addition on soil C storage were unrelated to the responses of cellulase activity (Fig. 1C). A model selection analysis (see Materials and Methods) confirmed that responses of soil C storage were best predicted by N-induced changes in LME activity over a broad range of climate factors, vegetation and soil types, and N application methods (Fig. 2). The response of LME activity also explained more variation in the response of soil C compared to a wide range of additional factors considered in the analysis (table S2; these factors were reported for only subsets of studies and so were analyzed individually).

Across the data set, N addition significantly decreased soil pH by 0.10 U (95% confidence interval, 0.02 to 0.17). Low soil pH can reduce decomposition rates and promote soil C storage (21, 23, 24). Thus, for the subset of studies reporting soil pH, we repeated our model selection procedure, including soil pH and treatment effects on soil pH as predictors. Responses of LME activity remained the most essential predictor of the effects of N addition on soil C storage (fig. S5). In addition, N addition also significantly increased the soil recalcitrant C pool by 22.7% and the proportion of recalcitrant C to total soil C storage by 9.2% (Fig. 3).

图 1.12　研究结果示例 2[15]

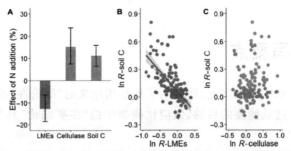

Fig. 1. Effects of N addition on LME activity, cellulase activity, and soil C storage (**A**). Relationship between the responses (ln R) of soil C storage to N addition and the response of LME activity (**B**) and cellulase activity (**C**). Error bars represent 95% confidence intervals; $n = 146$ in each panel. A negative relationship was found between the response of LME activity and the response of soil C storage [coefficient of determination (r^2) = 0.404, $P < 0.001$]. The light gray area indicates the confidence interval around the regression line. No significant relationship was found between the response of cellulase activity and the response of soil C storage ($r^2 = 0.008$, $P = 0.295$).

图 1.12（续）

图表较多时，可以考虑将部分图片放在补充材料中。关于图表的制作方法，参见第 10、11 章。

尽管结果部分是全论文最重要的部分，但它可能是最短的部分。例如在一篇 13 页的论文中，结果部分只占半页。

1.11 讨论撰写要点

如果说研究结果必须客观呈现的话，那么讨论部分则可由作者"自由发挥"。在讨论部分可以针对前文呈现的试验规律给出合理的机制揭示，也可以与前人的方法、已有的认识等进行对比分析，还可以说明论文中尚存在的缺陷或不足（图 1.13）。

（1）讨论应立足论文中的方法和结论来展开，而不是对方法和结论的简单重复。高质量的讨论应该是对研究结果的一次升华，重在揭示现象背后的本质。讨论还应回应引言中提出的问题，即是否全面或部分实现了引言最后一段提出的研究目标。

（2）讨论中不必羞于自夸，应客观评论自己工作的影响力，以及任何可能的工程应用；还要切忌试图隐藏论文的局限性。作者指出自己研究工作中存在的不足及其原因，不仅显示了论文作者对自己的研究工作有着全面的认识，还将有助于读者跟踪研究。

（3）讨论部分多用一般现在时态。

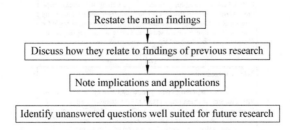

图 1.13　讨论逻辑结构示例

1.12 结论撰写要点

在研究结果和讨论的基础之上,简要明了地给出论文的"结论性"的认识。在此节中,通常无须总结全文——这是摘要的目的,但应比摘要中的"主要结论"具体。结论通常采用一般现在时态(图1.14)。

5. Conclusion
We conclude the following points:
1. Choice of reaction order has little effect (<0.3 orders of magnitude) on field-based estimates of time-integrated and time-averaged metamorphic fluid fluxes and metamorphic fluid flow durations based on reaction progress data.
2. Reaction order must be known for robust field-based determination of time-averaged net reaction rates based on reaction progress data. Underestimation of this term by more than 3 orders of magnitude can arise from assuming linear reaction kinetics.
3. Order of magnitude discrepancies between field-based and laboratory-based estimates of time-averaged net reaction rates can be explained by different reaction orders between laboratory experiments and natural metamorphic systems. In addition, porosity reduction coupled with reaction progress and/or nonlinear dependence of diffusivity on porosity can also contribute to this discrepancy.
4. Parameterization of metamorphic fluid flow is limited to time-averaged values which fail to account for the possibility that metamorphism occurs in short-lived pulses during longer time periods of metamorphic quiescence.

图 1.14 结论示例[11]

1.13 致谢撰写要点

致谢部分应提及所有对论文中的工作给予经费支持和技术支持的机构和个人。对于提供经费支持的研究机构,应注明合同编号;对于提供技术支持的个人,应注明其姓名、工作单位,以及具体的帮助内容。经过审稿后,也可在致谢中对期刊主编、审稿人提供的建设性意见表示感谢。

(1) 致谢中不建议包含"wish to""would like to"等类似表达。
(2) 致谢通常采用一般现在时(图1.15)。

Acknowledgments Financial support from the Bolin Center for Climate Research at Stockholm University is acknowledged. Hildred Crill is thanked for helpful comments on the writing of this manuscript. Prof. Alasdair Skelton from Stockholm University and Dr. Lanru Jing from the Royal Institute of Technology (KTH) Sweden are thanked for constructive comments and discussion. Two anonymous reviewers are also greatly acknowledged for their constructive comments.

(a)[1]

Acknowledgement
This work is supported by the jointly supported by the National Key R&D Program of China (2018YFC0407005), National Natural Science Foundation of China (51739006, 51779123, 51775295). The editor Prof Hsein Juang and the two anonymous reviewers are thanked for their constructive comments. The experimental data can be downloaded at: https://cloud.tsinghua.edu.cn/d/e5207512895845829f3e/.

(b)[20]

图 1.15 致谢示例

1.14 附录撰写要点

附录中的内容应与正文有紧密的相关性,且应是对正文内容的重要补充,比如原始数据、详细的公式推导过程等。

(1) 当存在多条目录时,附录应按"A,B,C,…"的形式进行编号。附录中包含的公式、图表等应与正文分开单独编号。

(2) 附录的时态应与正文中相应的内容保持一致。

1.15 参考文献撰写要点

参考文献通常包括与论文工作紧密相关的且已公开发表的各类论文、专著、报告、互联网资料等。正文中提到的参考文献,一定要在参考文献列表中提供完整的文献引用信息,并且所有列在文后的参考文献也一定要在正文中引用。

(1) 正文中引用参考文献的两种方式:顺序编码制(即提供文献序号)[图 1.16(a)]和作者-年代制(即提供作者姓氏+年份)[图 1.16(b)]。当作者为 2 人以下时,需把作者都列出;当作者为 3 人以上时,只需要列出第一作者并在其后加"等"字。

(2) 参考文献也有两种列表方式:按在正文中的引用顺序依次列出,或按作者姓氏首字母排序,或是这两种方式的结合(图 1.17)。

[11,12]. Also mechanical effects on solute migration in complex fracture systems have been addressed [13–15], but the feedback of chemically mediated changes on contaminant transport in fractured rocks has not yet been attempted.

(a) 顺序编码制(即提供文献序号)[14]

and both N_{Pe} and N_D are made nondimensionless by the arbitrary length scale, h. If D and ϕ are known, the time-averaged flux ($v\phi$) and the duration (t) of metamorphic fluid flow and k_{eff} can be directly obtained from inverse modeling based on reaction progress data. Previous estimates of $v\phi t$ have been summarized by Zhao and Skelton [2013] and range from 10^2 to 10^7 m^3 m^{-2} [e.g., Ferry and Gerdes, 1998; Zack and John, 2007; Winslow and Ferry, 2012]. Previous estimates of t range from 10^5 to 10^7 years [e.g., Christensen et al., 1989; Ague and Baxter, 2007]. Based on the above estimates, $v\phi$ generally ranges between 10^{-13} and 10^{-6} m s^{-1}.

(b) 作者-年代制(即提供作者姓氏+年份)[11]

图 1.16 正文引用参考文献示例

REFERENCES
1. Bear J, Tsang CF, Marsily DG. Flow and contaminant transport in fracturedrock. Academic Press Inc.: San Diego, 1993.
2. Gausl AP, André L, Lions J, Jacquemet N, Durst P, Czernichowski-Lauriol I, Azaroual M. Geochemical and solute transport modelling for CO2 storage, what to expect from it? *International Journal of Greenhouse Gas Control* 2008; **2**:605–625.
3. Zhao C, Hobbs BE, Ord A. Fundamentals of Computational Geoscience: Numerical Methods and Algorithms. Springer: Berlin, 2009.
4. Zhao C, Reid LB, Regenauer-Lieb K. Some fundamental issues in computational hydrodynamics of mineralization: A review. *J Geochem Explor* 2012; **112**:21–534.
5. Schwartz FW, Smith L, Crowe AS. A stochastic analysis of macroscopic dispersion in fractured media. *Water Resource Research* 1983; **19**:1253–1265.
6. Smith L, Schwartz FW. An analysis of the influence of fracture geometry on mass transport in fractured media. *Water Resource Research* 1984; **20**:1241–1252.
7. Cacas MC, Ledoux E, de Marsily G, Barbreau A, Calmels P, Gaillard B, Margritta R. Modeling fracture flow with a stochastic discrete fracture network: Calibration and validation: 2. The transport model. *Water Resource Research* 1990; **26**:491–500.
8. Moreno L, Neretnieks I. Fluid flow and solute transport in a network of channels. *J Contam Hydrol* 1993; **14**:163–192.
9. Bodin J, Delay F, de Marsily G. Solute transport in a single fracture with negligible matrix permeability: 1. Fundamental mechanisms. *Hydrogeology Journal* 2003; **11**:418–433.

(a) 按在正文中的引用顺序依次列出

图 1.17 参考文献列表示例[21]

References

Archie, G.E., 1942. The electrical resistivity log as an aid in determining some reservoir characteristics. Petroleum Technology 146, 54–62.
Bandis, S., Lumsden, A.C., Barton, N.R., 1983. Fundamentals of rock joint information. International Journal of Rock Mechanics and Mining Sciences & Geomechanics Abstracts 20, 249–268.
Barton, N.R., Bandis, S., Bakhtar, K., 1985. Strength, deformation and conductivity coupling of rock joints. International Journal of Rock Mechanics and Mining Sciences & Geomechanics Abstracts 22, 121–140.
Bear, J., 1972. Dynamics of Fluids in Porous Media. Elsevier, New York, 764 pp.
Berkowitz, B., Zhou, Z., 1996. Reactive solute transport in a single fracture. Water Resource Research 32 (4), 901–913.
Bodin, J., Delay, F., de Marsily, G., 2003a. Solute transport in a single fracture with negligible matrix permeability: 1. Fundamental mechanisms. Hydrogeology Journal 11, 418–433.
Bodin, J., Delay, F., de Marsily, G., 2003b. Solute transport in a single fracture with negligible matrix permeability: 2. Mathematical formalism. Hydrogeology Journal 11, 434–454.
Carslaw, H.S., Jaeger, J.C., 1959. Conduction of Heat in Solids, New York, 2nd ed. Oxford University Press, 510 pp.
Cook, N.G.W., 1992. Natural joints in rock: mechanical, hydraulic and seismic behaivor and properties under normal stress. International Journal of Rock Mechanics and Mining Sciences & Geomechanics Abstracts 29 (3), 198–223.
Cormenzana, I., 2000. Transport of a two-member decay chain in a single fracture: simplified analytical solution for two radionuclides with the same transport properties. Water Resource Research 36 (5), 1339–1346.
Danckwerts, P.V., 1953. Continuous flow systems: distribution of residence times. Chemical Engineering Science 2, 1–13.

(b) 按作者姓氏首字母排序[22]

References

[1] Adler PM, Thovert JF. Fractures and fracture networks. Dordrecht, Netherlands: Kluwer; 1999.
[2] Bai M, Elsworth D. Modeling of subsidence and stress-dependent hydraulic conductivity for intact and fractured porous media. Rock Mech Rock Eng 1994;27(4):209–34.
[3] Bai M, Meng F, Elsworth D, Roegiers JC. Analysis of stress-dependent permeability in nonorthogonal flow and deformation fields. Rock Mech Rock Eng 1999;32(3):195–219.
[4] Baghbanan A, Jing L. Stress effects on permeability in fractured rock masses with correlated fracture length and aperture. Int J Rock Mech Min Sci 2008;45(8):1320–34.
[5] Bear J, Tsang CF, Marsily De G. Flow and contaminant transport in fractured rock. San Diego: Academic Press Inc.; 1993.
[6] Berkowitz B. Characterizing flow and transport in fractured geological media: a review. Adv Water Resour 2002;25(8):3861–84.

(c) 两种方式的结合[23]

图 1.17（续）

（3）某些期刊要求在参考文献列表中将期刊名称按简写形式列出[图 1.17（c）]，期刊的简写形式可以从以下网站查询(http://cassi.cas.org/search.jsp)。

1.16 补充材料撰写要点

补充材料是正文中研究方法和研究结果的重要补充。补充材料的形式多种多样,包括文字、图表、照片、视频、软件和案例等。补充材料与附录的功能较为相似,具体撰写原则可以参考附录。

1.17 论文写作流程

论文写作过程如同盖楼(图 1.18)。撰写正文之前,应先草拟论文的标题和提纲。研究工作完成后,首先应将研究结果整理成图或表的形式,这是论文的基础。逐个给图表添加文字解释,即可构成研究结果和讨论的部分文字。再按研究方法、研究结果、讨论逻辑顺序完成论文核心部分的撰写。之后再对引言和结论部分进行完善或总结,并对参考文献进行认真核对。最后确定摘要、亮点、标题、关键词和致谢等必备部分。如需附录或补充材料,可以按要求撰写。

高效写作的若干建议:
(1) 时刻牢记论文撰写的黄金法则"Keep It Short and Simple"(KISS)(保持精简)。
(2) 研究工作过程中同时进行论文的写作,比如引言、方法等部分。
(3) 反复修改再投稿,切忌匆忙投稿。

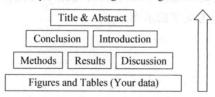

图 1.18 论文写作流程图

第 2 章　图表制作

图表是呈现复杂研究结果的最直接手段，故资深学者在阅读正文之前通常会快速浏览图表以对论文研究结果有初步的了解。学术论文对高质量图表的基本要求是确保读者仅通过图表及其标题即可理解学术论文的核心研究结果。此外，高质量的图片既可以吸引读者阅读全文，也可以体现论文作者的学术素养。本章重点介绍论文学术图表的制作原则，并辅以示例对关键注意事项进行说明。关于图表的具体制作流程及软件使用说明将在第 10 章、第 11 章中介绍。

2.1　表格 vs 图片

选择何种方式呈现研究结果的主要原则是：作者是想向读者传递精确的数值，还是向读者传递趋势或规律（图 2.1）。如果数据具有明显的趋势，建议采用图片来直观呈现。如果数据并无明显的趋势，或者数据本身就是重要的研究结果，则建议采用表格来准确呈现。当然，表格和图片的信息可能存在部分重复展示的情况（图 2.2）。此外，还应避免过度使用图表的情况，即相较于图表，有时采用文字来呈现研究结果可能会更加高效（图 2.3）。图 2.4 提供了一个将表格内容转换为文字的示例。

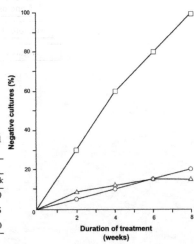

图 2.1　同一研究结果不同呈现方式[24]

第 2 章　图表制作

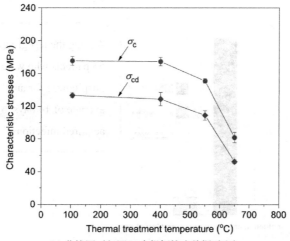

(a) 曲线图（与图(b)中粗框线内数据对应）

Group no.	Specimen no.	T_t (℃)	σ_c (MPa)	σ_{cd} (MPa)	σ_{cd}/σ_c	$\varepsilon_{1,c}$ (%)	$\varepsilon_{3,c}$ (%)	$\varepsilon_{v,c}$ (%)	$\varepsilon_{1,cd}$ (%)	$\varepsilon_{3,cd}$ (%)	$\varepsilon_{v,cd}$ (%)	E (GPa)	ν
1	BS28-1	105	179.21	133.86	0.75	0.3294	−0.2207	−0.1121	0.2446	−0.0686	0.1075	52.78	0.23
	BS28-2		178.06	136.23	0.76	0.3190	−0.1448	0.0294	0.2520	−0.0601	0.1319	51.31	0.19
	BS28-3		169.36	130.07	0.77	0.3042	−0.1583	−0.0124	0.2310	−0.0605	0.1100	54.21	0.21
	Mean		**175.54**	**133.39**	**0.76**	**0.3175**	**−0.1746**	**−0.0317**	**0.2425**	**−0.0631**	**0.1165**	**52.77**	**0.21**
	SD		5.39	3.11	0.01	0.0127	0.0405	0.0727	0.0107	0.0048	0.0134	1.45	0.02
2	BS28-4	400	177.67	133.80	0.75	0.4262	−0.1802	0.0658	0.3404	−0.0509	0.2385	34.39	0.08
	BS28-5		168.55	119.60	0.71	0.4084	−0.1536	0.1011	0.3178	−0.0494	0.2190	33.12	0.09
	BS28-6		177.50	133.12	0.75	0.4582	−0.1457	0.1668	0.3648	−0.0510	0.2629	31.44	0.08
	Mean		**174.57**	**128.84**	**0.74**	**0.4309**	**−0.1560**	**0.1112**	**0.3410**	**−0.0504**	**0.2401**	**32.98**	**0.08**
	SD		5.22	8.01	0.02	0.0252	0.0181	0.0513	0.0235	0.0009	0.0220	1.48	0.01
3	BS28-7	550	150.39	102.44	0.68	0.6798	−0.3291	0.5526	0.5206	−0.0160	0.5206	15.56	−0.02
	BS28-8		148.49	113.65	0.77	0.6920	−0.0855	0.5210	0.5956	0.0096	0.6149	14.91	−0.08
	BS28-9		153.81	110.51	0.72	0.6842	−0.1569	0.3703	0.5750	0.0079	0.5908	15.63	−0.08
	Mean		**150.90**	**108.87**	**0.72**	**0.6853**	**−0.1393**	**0.4068**	**0.5744**	**0.0005**	**0.5754**	**15.37**	**−0.06**
	SD		2.70	5.78	0.05	0.0062	0.0475	0.1010	0.0215	0.0143	0.0490	0.40	0.03
4	BS28-10	650	88.26	53.84	0.61	1.2486	−0.1321	0.9844	1.0830	0.1158	1.3146	4.27	−0.13
	BS28-11		81.65	53.69	0.66	1.3040	−0.1138	1.0764	1.1482	0.1234	1.3949	3.77	−0.13
	BS28-12		75.31	49.50	0.66	1.3020	−0.1865	0.9290	1.1256	0.1086	1.3427	3.60	−0.13
	Mean		**81.74**	**52.34**	**0.64**	**1.2849**	**−0.1441**	**0.9966**	**1.1189**	**0.1159**	**1.3507**	**3.88**	**−0.13**
	SD		6.48	2.46	0.03	0.0314	0.0378	0.0745	0.0331	0.0074	0.0407	0.35	0

T_t: thermal treatment temperature; σ_c: peak stress; σ_{cd}: crack damage stress; $\varepsilon_{1,c}$: axial strain that corresponds to the peak stress; $\varepsilon_{3,c}$: lateral strain that corresponds to the peak stress; $\varepsilon_{v,c}$: volumetric strain that corresponds to the peak stress; $\varepsilon_{1,cd}$: axial strain that corresponds to the crack damage stress; $\varepsilon_{3,cd}$: lateral strain that corresponds to the crack damage stress; $\varepsilon_{v,cd}$: volumetric strain that corresponds to the crack damage stress; E: Young's modulus; ν: Poisson's ratio. The mean and standard deviation (SD) values of parameters for different groups are indicated by bold font.

(b) 试验数据

图 2.2　数据呈现形式示例[25]

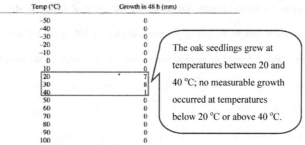

(a) 过度使用表格

图 2.3　过度使用图表示例[24]

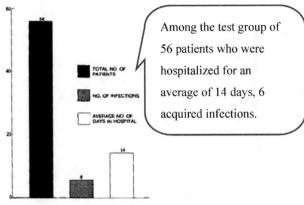

(b) 过度使用图片

图 2.3（续）

Table 2　Microscopic parameters of the numerical Beshan granite specimen

Type	Parameters (unit)	Value
Particle	Contact modulus (GPa)	51
	Ratio of particle shear to normal stiffness	1.0
	Contact friction coefficient	0.5
Parallel bond	Parallel bond radius multiplier	1.0
	Modulus (GPa)	35
	shear to normal stiffness	1.0
	Normal strength (mean + standard deviation in MPa)	104 + 41.6
	Shear strength (mean + standard deviation in MPa)	104 + 41.6

审稿人意见及作者回复：

Comment 4 from the Editor：Please delete Table 2, and describe this data in the text.

Response：Table 2 has been deleted, and the microscopic parameters are described in the text (lines 344-347).

将表格内容转换为如下文字：

> Based on the uniaxial compression test results and the initial thermal conductivity measurement data of the specimens preconditioned at 105 °C, a trial-and-error process was performed to calibrate the microscopic parameters of the numerical specimen, i.e., contact modulus of 51 GPa, ratio of particle shear to normal stiffness of 1.0, contact friction coefficient of 0.5, parallel bond radius multiplier of 1.0, parallel bond modulus of 35 GPa, ratio of shear to normal stiffness of 1.0 for parallel bonds, parallel bond normal strength of 104 ± 41.6 MPa and parallel shear strength of 104 ± 41.6 MPa.

图 2.4　根据审稿意见将表格内容转换为文字表达[25]

2.2 表格制作要点

表格通常用于呈现大量数据,作者应仔细设计表格,以方便读者清楚地获取数据。设计制作表格应遵循的原则有:

(1) 通常将自变量置于表格左侧竖向排列,而将因变量置于表格顶端横向排列。提供清晰简明的标题、图例、单位等重要信息。

(2) 文字通常左对齐,而数字通常右对齐。列和行之间有足够的间距,字体类型和大小清晰易读。

(3) 表格应该按文字中提及的顺序依次编号。当多个表格呈现相似的数据时,应采用相似的表格形式(图 2.5)。

Table 1. Best Fit Parameters for Infiltration of CO_2-Bearing Fluid Through Selected Metabasaltic Sills in the Dalradian of the Southwest Highlands in Scotland

Reaction Order n	Time-Averaged Flux $\log(v\phi)$ (m/s)	Duration $\log(t)$ (year)	Effective Net Rate Constant $\log(k_{eff})$ (s^{-1})	Time-Integrated Flux $\log(v\phi t)$ (m)	R^2
(a) Port Cill Maluaig (High N_{Pe} and High N_D; N_{Pe}/h = 5.0 m^{-1}, N_D/h = 1.7 m^{-1})					
1	−10.12	4.76	−10.19	2.14	0.90
1.6	−10.16	4.83	−9.75	2.17	0.90
2	−10.13	4.82	−9.24	2.18	0.91
2.7	−10.15	4.86	−8.17	2.21	0.91
(b) Jura 11 (Low N_{Pe} and High N_D; N_{Pe}/h = 0.01 m^{-1}, N_D/h = 8.7 m^{-1})					
1	−11.35	5.23	−10.44	1.39	0.79
1.6	−11.41	5.31	−9.55	1.40	0.79
2	−11.43	5.34	−8.97	1.41	0.79
2.7	−11.46	5.39	−7.97	1.43	0.80
(c) Loch Stornoway E (High N_{Pe} and Low N_D; N_{Pe}/h = 13.3 m^{-1}, N_D/h = 3.4 m^{-1})					
1	−8.58	2.29	−7.52	1.21	0.96
1.6	−8.50	2.24	−6.35	1.24	0.96
2	−8.49	2.25	−5.60	1.26	0.97
2.7	−8.48	2.26	−4.35	1.28	0.97
(d) Port Cill Maluaig C (Low N_{Pe} and Low N_D; N_{Pe}/h = 0.1 m^{-1}, N_D/h = 0.3 m^{-1})					
1	−11.33	4.67	−10.78	0.84	0.85
1.6	−11.45	4.82	−10.61	0.87	0.82
2	−11.49	4.89	−10.48	0.91	0.81
2.7	−11.53	5.00	−10.24	0.97	0.79

图 2.5　表格示例[26]

2.3 图片

图片是学术论文研究成果形象化呈现的理想形式,其类型有数据图、示意图、图像、地图和流程图等。以下分别给出设计制作各类图片应遵循的原则。

(1) 数据图:通常用于呈现两个或多个变量之间的函数或统计关系。数据图重在呈现变量之间的宏观联系,而非单个数据点的详细信息。在数据图中应注明所有坐标轴名称和单位以及所有曲线和数据集对应的变量,并使用清晰的字体和字号,避免使用浅黄、浅绿等颜色(图 2.6)。

(2) 示意图:通常用于说明研究结果背后的物理机制或研究方法的基本原理。在示意

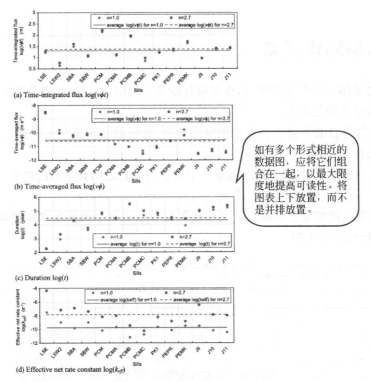

图 2.6 数据图示例[11]

图中应配合采用符号或文字标注关键信息(图 2.7)。

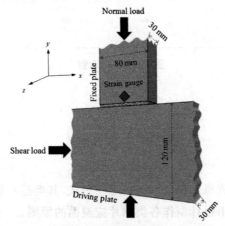

(a) 试验装置[27]

图 2.7 示意图示例

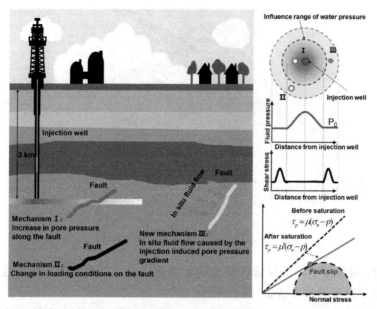

(b) 物理机制[20]

图 2.7（续）

（3）图像：通常用于呈现研究对象及其在特定条件下的演化过程。首先应确保图像的清晰度，还应明确给出比例尺、关键信息标记、颜色和符号的意义等重要信息（图 2.8）。

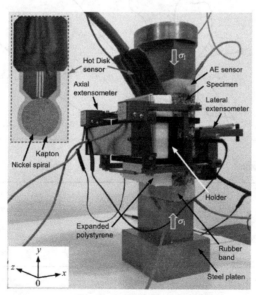

(a) 研究对象与试验装置[25]

图 2.8　图像示例

(b) 微观试验结果[28]

图 2.8(续)

（4）地图：通常用于标识现场研究工作的地理位置。在地图中应标明经度和纬度、比例尺、图例、项目位置等重要信息（图 2.9）。

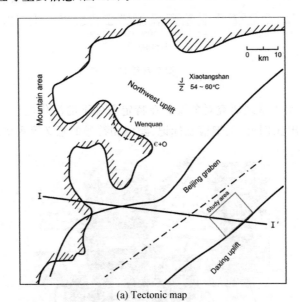

图 2.9　地图示例[29]

（5）流程图：通常用于梳理研究过程的逻辑关系。在流程图中应采用合理的图框对应每一步操作，并标明操作步骤之间的衔接关系（图2.10）。

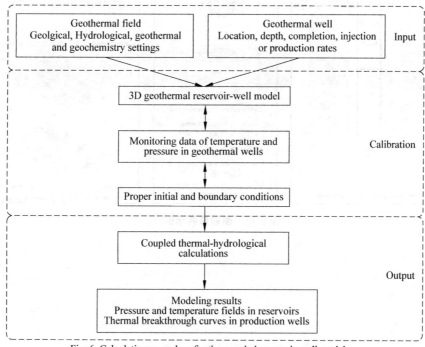

图2.10　流程图示例[30]

2.4　图形摘要

图形摘要（graphical abstract）是学术论文研究成果高度形象化的概括形式。虽然图形摘要可以是正文中结论性的图片，但作者通常会专门设计图形摘要，以方便读者一目了然地理解论文的核心学术贡献。图形摘要通常以在线形式发表，一般不出现在正式印刷版的论文中。设计制作图形摘要应遵循的主要原则有：

（1）图形摘要主要用于描述一个重要流程或是重要方法，故应该是一单独且清晰的图形文件。图形摘要应具有醒目的开始和结束，建议从上至下或从左至右进行设计，还应尽量减少分散注意力的元素。

（2）图形摘要仍须辅助必要的文字说明，建议采用字号足够大的 Times New Roman、Arial 等字体（图2.11）。

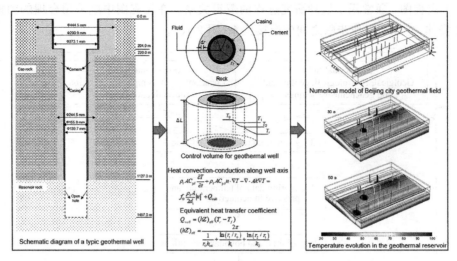

图 2.11　图形摘要示例[30]

第 3 章 写作范式

内容、结构和语言是学术论文写作的三要素。研究成果取决于作者的学术洞察力与研究水平,这些不在本书讨论范围之内。学术论文在宏观上应严格按照第 1 章介绍的 IMRD 组织架构,而在语言层面则允许作者形成自己独特的行文风格。用语是否准确得体将是本章要重点讨论的内容,这也是论文作者学术素养的重要体现。英语是当前科技论文写作的通用语言,广大母语非英语的科研工作者在撰写科技论文时应努力做到兼顾文章内容、结构和语言。

3.1 学术论文风格

与文学作品不同,学术论文的目的是准确客观地向读者展示研究成果,故学术论文在行文方面的最重要原则就是"准确和简明",即用简洁的语言将研究成果客观地表达出来,使读者易于理解文章内容,促进更深更广的学术交流。学术论文写作推荐采用陈述句,简单的"主谓宾"或"主系表"结构最适用于学术论文的表达需求。一方面,便于母语非英语的学者理解,扩大论文的受众群体;另一方面,此结构的语法最为简明,能极大地避免论文出现语法错误。建议在领域内主流期刊上检索出几篇与自己课题类似的文章,归纳文章各部分的常用表达方式和词汇,以供自己写作时参考。

强调学术论文要易于读者理解,并非指科研工作肤浅没有深度,而是指论文的语言要平实、易于理解。学术论文的读者群不仅包括同专业的学者,还会有来自不同领域的工程技术人员,而且不分国界,所以为方便各国读者都能准确理解论文内容、扩大学术交流的深度与广度,必须在论文的可读性上下功夫。同时论文可读性强也有利于审稿人给出更充分、客观和公正的审稿意见。总而言之,论文的语音表达应该做到直截了当、开门见山、不绕弯子,即紧紧把握住"平实"二字。

学术论文在语言表达方面的具体注意事项有:

(1) 句子应尽量结构简单,且不宜过长。多用肯定句。

例 1: The gouge particle rolled at the beginning period of shear process, accompanied by micro-cracking inside the gouge particle and surface erosion, and then micro-cracks propagated through the gouge particle, i.e., full breakage occurred.

将长句拆为短句,易于读者理解。

The gouge particle still rolled at the beginning period of the shear process (before breakage), accompanied by micro-cracking inside the gouge particle and surface erosion. With increasing shear displacements, the micro-cracks fully propagated through the gouge particle and full breakage occurred.

例 2：The shear stress continuously kept fluctuating during the whole shear process, because breakage occurred continuously during the whole course.

将"先果后因"变为"先因后果"，同时删除不必要的重复。

Gouge breakage occurred continuously during the whole course, so the shear stress kept fluctuating.

例 3："It is not necessary to remove …"

为什么要用形式代词这种结构呢？如修改以下句子结构，将更便于读者理解。

"… need not be removed."

如果删繁就简可以把 need not 变成 needn't，但学术论文中不推荐"n't"等缩写。

（2）用词一致性：对同一研究对象，要自始至终用一个词来描述，注意学术论文不是展示词汇量的舞台。

例：crack、fissure、flaw、fracture 等词都有裂缝或裂隙的意思，但在土木工程不同领域的专业术语中有不同的意义，比如在岩石力学领域通常特指不同尺度的裂缝（图 3.1）。

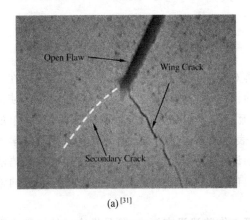
(a)[31]

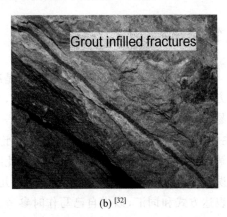
(b)[32]

图 3.1　不同尺度岩石裂缝的特定用词（fracture 与 fissure 意思相近，但在学术论文中 fracture 用得更多些）

（3）选词原则：尽量选用意思较为单一的词，母语非英语的人要理解一词多义较难。避免使用俚语、成语、谚语等。

例："much easier"较"a good deal easier"易于理解。

"should not be used repeatedly"较"do not bear repeated use"易于理解。

3.2　语态

主动或被动语态的选择主要取决于作者要强调的内容，当然也与作者的写作风格有关系。表 3.1 简单对比了两种语态的特点，并给出了学术论文中选择语态的主要原则。目前，部分期刊社在作者指南中明确建议多采用主动语态。但是，被动语态的非人格色彩，确实能够使论文表达更加客观化。具体到学术论文各主要章节，主动语态较多地出现在引言和讨论中,而被动语态在研究方法和研究结果中采用的概率更高些。

表 3.1 主动语态与被动语态的对比选择

语态类型	主动语态	被动语态
特点	简明扼要,开门见山	语言结构为逆序
优劣	语言结构为顺序,条理清晰,更适合科技论文简明的写作原则	可以省略动作的发出者,故可突出动作本身或动作的接受者,即研究方法和研究对象
总结	学术论文写作多采用主动语态;当动作的发出者无法确定,或要凸显动作的接受者或动作本身,则使用被动语态,即被动语态尽量少用	

例 1:论文各主要章节语态示例

These studies contributed greatly to the understanding of ...（引言或讨论中可采用的主动语态句式）

In this study, we conducted ... using ..., and we focused on ...（引言或讨论中可采用的主动语态句式）

We examined the... data sets using surface amplitude-based roughness parameters including...（研究方法中可采用的主动语态句式）

Two... samples were prepared and tested in this study.（研究方法中可采用的被动语态句式）

We found that ... is best fit by a two-term power law model.（研究结果中可采用的主动语态句式）

The lower portion of each curve can be fitted by a linear relation, suggesting...（研究结果中可采用的被动语态句式）

例 2:当以物体做主语时,应避免如下错误使用被动语态的情况。

When ..., fracture aperture gently closed at the initial period. This was resulted from ...（错误）

When ..., fracture aperture gently closed at the initial period. This resulted from ...（正确）

As normal stress increased, more micro-cracks were developed in the shallow zones of fracture surfaces.（错误）

As normal stress increased, more micro-cracks developed in the shallow zones of fracture surfaces.（正确）

3.3 时态

尽管在第 1 章中已经给出了论文各主要章节建议采用的时态,在本节中还将详细介绍时态使用的基本原则。采用过去时描述之前发生的情况,比如:做了什么,前人报道了什么,研究中发生了什么等;采用现在时描述一般事实,比如:结论(包括作者或前人得出的结论)和客观描述(比如图表所表达的信息);采用未来时态展望未来的工作(表 3.2)。因此,引言、讨论等章节通常应采用现在时,而摘要、方法、材料、结论等章节通常是应采用过去时,因为前者主要是叙述前人的成果(已经发表、业内认证、客观事实),后者主要是论述自己的成果(尚未发表、未经认证、过去的工作)。

表 3.2 时态示例

时态	语境	示例
过去时	已完成的工作	We collected rock samples from...(研究方法)
		Zhao et al. (2017) developed a new rotary shear test apparatus that allows...(引言)
	已观察到的现象	For both experiments, the total numbers of contact patches decreased with increasing...(研究结果)
		The aspect ratio of contact patches of the ... test ranged from 1.0 to 8.4.(研究结果)
	已报道的工作	Recently, Petrovitch et al. (2013) demonstrated numerically that...(引言)
		Many studies suggested that frictional behavior is related to...(引言)
现在时	结论	Volumetric flow rates are controlled by the size and spatial distribution of the apertures of the void space.(引言)
		This scale-dependent behavior is partly attributed to ... (Barton and Choubey 1977).(引言)
	客观描述	Figure 12 presents the sieve analysis of crushed gouge pieces with increasing normal stresses.
		Figure 14b shows the fracture closure with shear displacement, which exhibits that the model of ... had the largest aperture closure.
将来时	展望未来	Failure to do so will render the resulting roughness parameters incomparable.
		Not only gouge material in the fracture can influence the mechanical behavior of fractures, but also their evolution can be expected to significantly change the fluid transmissivity, flow pattern and solute transport as a result, which will be studied in future research.

在描述研究结果时,经常涉及现在时和过去时的交叉使用,比如以下情况:

(1)"Smith (2001) showed that ... inhibits ...",其中 showed 指过去发生的事情,因此是过去时;而 inhibits 则是指该论文发现的普遍规律,故采用现在时。

(2)"Figure 1 shows that ... inhibited ...",其中 shows 指图 1 呈现的内容,因此是现在时;而 inhibited 则是特指图 1 支撑的具体研究结果,故采用过去时。

(3)"Significant amounts of ... were isolated. This indicates that ... is ...",其中 were isolated 特指某次研究中的结果,而 indicates that ... is 则是指基于该结果推论得出的一般性规律。

3.4 用词

准确是学术论文写作的最基本要求之一,而用词的准确性是实现该要求的重要一环,也是写作中的一大难点。对于母语非英语的作者,必须弄清每个单词的真正含义和使用环境,不能仅根据字面直译结果盲目使用单词。

(1) 动词:用于描述动作、状态或事件。在学术论文写作中应采用动词明确行为或过程,并使句子简洁,不建议将动作状态或行为掩埋在名词中。

例:推荐"The temperature reduced the rock compressive strength significantly.",而不建议"The temperature produced a significant reduction in rock compressive strength."。

(2) 名词:用于描述事物或现象。在学术论文写作中应注意区分名词的单复数形式。

例:The rock data are sorted in Table 1. 其中 data 是指特定的数值结果,因此应将其视为复数名词。而在信息科学中,data 通常指要访问、存储、处理的大量信息,与 information 相类似,按单数形式处理。

research 为不可数名词,而 study 为可数名词。

(3) 形容词:用于描述事物或现象(名词)的属性或特征。在学术论文写作中应避免采用过度修饰。

例:New invention,其中 new 属于<u>重复修饰</u>。

(4) 副词:用于描述动作、状态或事件的程度与频率等。与形容词一样,在学术论文写作中应避免采用过度修饰或模糊修饰。

例:This method illustrates the frequency of very high-energy collisions. 其中,very 属于模糊修饰,在学术论文写作中应慎用。同样应该慎用的副词还有 quite。

(5) 数词:用于描述事物的个数或变量的数值。在学术论文写作中采用数词应遵循以下原则,对于整数 1~9 用单词形式,而其他所有数字都采用数字形式;当数字位于句首时采用单词形式;数字与单位之间应插入一个空格。

例:A rejection rate of 70 m^3/h is too high.

(6) 冠词:包括定冠词 the 和不定冠词 a、an,定冠词用于指代特定事物;不定冠词用于修饰非特定事情(表 3.3)。

表 3.3 冠词使用语境与示例

冠词	使用语境	示 例
零冠词	大多数单数专有名词之前 包含其他限定词 一般性地描述非专有的事物	Europe this study Water is essential...
定冠词	世上独一无二的某个事物 人尽皆知的名词或事物 紧跟一段限定性描述的名词 由修饰语特殊限定的名词 在上文已被提及的名词	The university is... The north The data points that... The data points of...

续表

冠词	使用语境	示 例
不定冠词	若没有定冠词或其他指代词(如this, our)修饰,可数名词单数前加不定冠词 指代任一团体或单一成员的名词 首次提出一个无明确指代的可数名词单数	A glacier is a body of ice. This is an introductory sentence.

(7) 介词:是在名词、代词或名词短语之前使用的一个单词或一组单词,用于指示方向、时间、位置、位置、空间关系或引入对象等。

例:The effects of microparameters used in the particle mechanics models on the simulation of gouge behaviors were also investigated by sensitivity analysis. 建议将 by 改为 though 或 using。介词 by 后面跟的一般是人,而 through 或 going 后面跟的一般是工具。

3.5 常见错误

写作方式囊括了英语语法的所有方面,本章前几节介绍了在学术论文写作中应注意的语法原则,由于空间所限,这里不可能穷尽所有的语法细节,以下仅罗列学术论文中常见的语法错误供读者参考。

(1) 悬垂修饰语。这种语法结构并不清晰,易产生歧义,应在学术论文写作中尽量避免悬垂修饰语。

例:Before adding the compound, it was determined that the solution's pH was 6.4. (错误)

此句中,it 无法在 solution 中添加 compound,如果仍采用被动语态,可以修改为:Before the addition of the compound, it was determined that the solution's pH was 6.4. (正确)

当然,更为简洁的表达方式是采用主动语态,如下所示:Before adding the compound, I determined that the solution's pH was 6.4. (正确)

(2) 可数名词单复数与主谓一致。一些名词拥有相同的单复数形式,如 literature、equipment、staff、faculty 等,名词单复数的使用根据数量确定即可,单复数变化规则在此不赘述。

主谓一致就是要谓语与主语的单复数保持一致,需明确两个问题:主语是什么? 主语的单复数如何确定? 比如:"10 g was added"和"10 g were added",主语显然是"10 g",前者将主语看作一个量度,一个整体,即单数,故谓语使用"was";后者将主语看作质量的数,强调"按克加",这个数"10"是复数,故谓语使用"were"。

(3) 缩略形式。虽然学术论文中不提倡使用缩略词,但如果某些词组被反复提及时,也可按以下原则适当使用缩略词:文章标题不使用缩略词,而在摘要与正文中,若一个长词或词组,需要反复使用多次,可以考虑缩略词;若某个长词或词组只出现一两次,不建议使用缩略词,若该长词或词组被频繁使用(3~6次),可使用缩略词。

写初稿时,先不使用缩略词。待初稿完成,进行复核时,留意频繁出现的词语,对符合缩略词使用条件的,可在后期进行替换。首次提及某词时,使用完整拼写,并在全称后用括号括起其缩略词形式,后续使用该词即可直接使用缩略词形式。

(4) amount vs number

amount 强调一个整体的量,故多用于修饰不可数名词,如 an amount of water;而 number 强调个体,故多用于修饰可数名词,如 a number of samples。

(5) like

like 不能用作连词,只能作介词(接宾语),在书面语中要表达上"像……"的意思建议采用"as",如 Even though under normal stress of 1.0 MPa, either gouge particles was not fully broken like single gouge case. 建议将 like 改为 as。

(6) varying

误把 varying 归在 vary 系列,认为与 various 等同根词有类似的意义,但 varying 实际是 changing 的意思,表示变化;而 various 是"多种多样的",表示一个事物的各个方面或各种类型,但这个事物往往不发生变化。

various concentrations(不同浓度);

varying concentrations(变化的浓度)。

(7) while

while 作"然而",最适合存在时间关系的情况,如果某个转折缺少时间的背景,用"whereas"更好。

(8) 冠词使用错误

Figure 2 shows the distribution of relative velocity on(the) surface of (the) main and splitter blades.

The software ××× is chosen to be a(the) 3D modeling tool in this study.

A theoretical method for calculating the inner flow-field in(a) centrifugal impeller with splitter blades and (an) investigation of the interactions between the main and splitter blades is presented in this paper.

(9) 形容词/副词使用不当

The results show that gouge particle ~~may~~ behave in two different ways under ~~small or large~~(low and high) normal stresses.

Under larger normal stress, gouge particles can be crushed in to a few major pieces and a large number of minor comminuted particles, accompanied by ~~severer~~(more severe) damage on fracture walls and fracture closure.

Gouge materials is usually generated from rock fracture surfaces undergoing compression and shear, and subsequently these gouge particlescan be further crushed into (even) smaller ones with ~~further~~(further) shear displacement or ~~increasing~~ normal stress.

(10) 介词使用不当

... can play a key role ~~on~~(in) fracture mechanical properties.

However, the progress ~~in~~(of) research on ... has been slow.

~~Different from~~(Unlike) previous studies, this research focused on ...

... proposed a scheme to replace the original particles that fulfill failure criterion ~~by~~(with) a set of smaller particles.

第 4 章 论文投稿

论文撰写完毕后,投稿与返修就成为论文发表的又一关键环节。在此过程中作者不仅会涉及与主编、编辑、审稿人的交流,而且还须根据审稿意见对论文进行修改,以保证论文能被顺利接受并发表。事实上,土木工程领域国际英文期刊的录用率普遍都低于25%左右,由此可见论文发表的难度。尽管高质量的研究成果和论文写作固然是学术论文发表的重要前提,但是投稿与返修环节的疏忽或失误也会导致论文不能顺利发表。本章将详细介绍论文投稿、审稿、返修、校样等过程中可能遇到的各种问题及正确合理的处置方法。

4.1 期刊选择

将完成的学术论文投稿至合适的期刊是保证论文顺利发表并被广泛阅读引用的重要前提。选择拟投稿的期刊一般应综合考虑以下因素:

(1) 专业:尽量选择与论文主题相符的期刊。每种期刊涵盖的主题一般都能从期刊的主页获取(图4.1)。此外,还应采用论文的关键词检索拟投期刊之前的论文,如有相关主题的论文,也说明期刊与论文的主题相一致。需要注意的是,期刊的主题也存在动态调整的可能。

(2) 声望:根据论文质量尽量选择领域内声望较好的期刊,即领域内的权威或主流期刊。经过长时间的积累,期刊的声望一般是领域内同行所公认的,而期刊的影响因子是期刊质量最直接的指标之一。学者在领域内的声望主要取决于其在领域内权威或主流期刊上发表论文的数量,而非学术论文的绝对数量。

(3) 读者:投稿前须了解拟投期刊的主要读者群,尽量选择读者群中本领域学者所占比重较大的期刊,这样文章的潜在读者才会多,日后文章被引量才会更高。

(4) 周期:这里主要指审稿周期,尽量选择审稿周期较短的期刊,以使论文早日见刊发表。除审稿周期外,还包括发表周期。土木工程领域国际英文期刊的审稿周期一般为2~6个月,而发表周期为4~12个月。

如果投稿至不合适的期刊,可能会有以下结果:①因与期刊主题不符而被直接退稿(图4.2);②因找不到合适的审稿人导致审稿周期变长或无法给出客观审稿意见;③被领域内从事相关研究的学者忽视,导致论文引用率较低。

Engineering Fracture Mechanics
An International Journal

Editors: A.R. Ingraffea, M. Kuna, X.Q. Feng

> View Editorial Board

> CiteScore: 3.46   Impact Factor: 2.908

Published in Affiliation with the European Structural Integrity Society

EFM covers a broad range of topics in fracture mechanics to be of interest and use to both researchers and practitioners. Contributions are welcome which address the fracture behavior of conventional engineering material systems as well as newly emerging material systems. Contributions on developments in the areas of mechanics and materials science strongly related to fracture mechanics are also welcome. Papers on fatigue are welcome if they treat the fatigue process using the methods of fracture mechanics.

The Editors especially solicit contributions which synthesize experimental and theoretical-computational studies yielding results with direct engineering significance.

图 4.1　*Engineering Fracture Mechanics* 期刊简介

（https://www.journals.elsevier.com/engineering-fracture-mechanics/）

Ref: ATE_2020_345
Title: Heat transfer in…
Journal: Applied Thermal Engineering

Dear Dr. (Author's name),

Thank you for submitting your manuscript to Applied Thermal Engineering. We regret to inform you that your manuscript does not fit within the scope of the journal and we are therefore returning it for you to explore alternative publication outlets. We are sorry to disappoint you with this decision and hope that you will be able to successfully submit your manuscript elsewhere.

Thank you for giving us the opportunity to consider your work.

Kind regards,
Dr. (Editor's name)
Deputy Editor-in-Chief
Applied Thermal Engineering

图 4.2　因与期刊主题不符而被直接退稿示例

4.2　论文投稿

确定拟投期刊后，应仔细阅读期刊的投稿须知，根据期刊要求准备好论文各要件，并修改论文格式。论文各章节的撰写要求已在第 1 章详细介绍过，在此不再赘述。一般按照期刊格式要求调整论文格式即可。目前多数期刊允许作者在首轮投稿时按"Your Paper Your Way"的形式准备论文草稿。即使如此，作者应在准备论文草稿时注意如下基本的格式要求：1.5~2倍行距；页边距不小于25 mm；添加行号。

多数期刊还要求在投稿时提交自述信(cover letter)。在自述信中，作者可以阐述文中研究工作的意义和创新点，还可以解释论文主题与期刊宗旨的契合度，以及可能感兴趣的读者群，从而有助于主编了解论文的主要学术贡献，并选择合适的审稿人。图4.3给出了两封自述信的范例，第一封供首次投稿参考，另一封供拒稿后重投参考。

Coverletter

[Your Name]
[Your Affiliation]
[Your Address]

[Journal title]
Date

Dear Porf./Dr. [Editor's name],

I/We am/are writing to submit an original manuscript entitled "[title of paper]" for consideration by [journal name].

In this manuscript, I/we report on / show that _____. This is significant because _____.[Please explain in your own words the significance and novelty of the work, the problem that is being addressed, and why the manuscript belongs in this journal. Do not simply insert your abstract into your cover letter! Briefly describe the research you are reporting in your paper, why it is important, and why you think the readership of the journal would be interested in it.]

We believe that this manuscript is appropriate for publication by [journal name] because it… [specific reference to the journal's Aims & Scope]. _____.

I/We confirm that this work is original and has not been published elsewhere, nor is it currently under consideration for publication elsewhere.

We have no conflicts of interest to disclose.

Thank you for your consideration of this manuscript.

Sincerely,
[Authors' names]
[Corresponding author's email]

(a) 首次投稿

图 4.3 投稿自述信示例

```
                                              Coverletter

                                                                    [Your Name]
                                                                 [Your Affiliation]
                                                                   [Your Address]

          [Journal title]
          Date

          Dear Porf./Dr. [Editor's name],

          Please find enclosed a paper for submission to EPSL entitled "[title of paper]".
          This paper is based on a paper which was previously submitted to EPSL
          (EPSL-D-12-00×××). The paper was reviewed by two anonymous reviewers, who
          were overall positive, but Reviewer 1# questioned the paper appropriateness for
          the journal, because its focus was on "presenting a new inverse modeling
          framework". You rejected our paper on this basis and after reading the reviewers
          comments, we agreed with your conclusion.

          Since then, we have rewritten our paper entirely, changing its emphasis from a
          study presenting a new inverse modeling framework, to a study which uses a new
          numerical model to address long standing and highly relevant questions pertaining
          to our understanding of…. In this new submission, we simultaneously
          calculated … based on field data without employing a steady (or quasi-stationary)
          state approximation. This has not been done before. We also calculated … and
          showed them to be comparable. We also showed that …. Finally, we examined the
          roles of parameters which were dependent on ….

          We fully understand that you will need to treat this paper as a new submission. If
          you are prepared to send this paper out for review, our preference would be for the
          paper to be sent back to the previous reviewers, but we leave this of course entirely
          to your discretion.

          Thank you for your consideration of this manuscript.

          Sincerely,
          [Authors' names]
          [Corresponding author's email]
```

(b) 拒稿后重投

图 4.3（续）

4.3 审稿

论文审稿周期由数周至数月不等，期间作者不可避免地需要与期刊编辑交流审稿事宜。对于多数期刊，主编（editor-in-chief）或副主编（associate editor）是论文接收与否的决策者，他们通常也是领域内颇具声望的学者。执行编辑（managing editors）一般不直接参与论文的审稿过程。论文被接受前的问题通常在主编或副主编的职权范围内，而被接受后的问题则由执行编辑负责处理。此外，还有文字编辑，主要负责按照期刊规定的格式、风格编辑论文,纠正文中语法、拼写、符号、用词等方面的错误。

多数期刊的审稿流程大同小异。首先是格式和主题审查:稿件投稿至期刊后,先由执行编辑或助理编辑对论文是否符合期刊格式要求进行审查。若格式不符,稿件会被立即退回修改(图4.4),直至满足要求。格式审查通过后,主编或副主编会审查稿件内容是否在期刊所涵盖的领域之内。若主编或副主编判断论文与期刊主题不符,通常会直接拒稿(图4.2)。遇到此类拒稿,建议作者改投其他合适的期刊。即便花费时间与主编或副主编争论与期刊主题是否相符,通常也不会改变主编或副主编的拒稿决定(图4.5)。

> Journal of Geophysical Research - Solid Earth: Your Submission Files: MS# 2013JB01051X
>
> Dear Dr. [Author's name],
>
> Thank you for your recent submission to Journal of Geophysical Research - Solid Earth ("PAPER TITLE", [Paper reference number]).However, we are unable to forward your paper to the Editor to start the peer review process until the following problems are corrected:
>
> — Please ADD line numbers to your article file. The line numbers should be continuous from the first page of your article file to the end.
>
> — Please add your Figure Caption to your Article File.
>
> — The citations below were run through our parser check which identifies citations that do not appear to be correct according to CrossRef (an online database of most peer-reviewed and scientifically indexed content).
>
> Citations that are from older books, theses or user manuals may be indicated below. Please double check to make sure that these citations are listed correctly.
>
> Please check the remaining citations to make sure that they are complete and up-to-date to the best of your knowledge. Some common errors include (but are not limited to) incomplete abbreviations and incorrect or missing DOI numbers. Thank you for taking the time to make the changes to your Reference List.
>
> * Balashov, V.N., Yardley, B.W.D., 1998 Modeling metamorphic fluid flow with reaction-compaction-permeability feedbacks. Am. J. Sci. 298, 441-470.
>
> If you have any questions about how to complete your submission, please contact me by replying to this e-mail message or click on the links below for additional online help.
>
> Sincerely,
> [Name]
> Editor's Assistant, Journal of Geophysical Research - Solid Earth

图4.4　因不符合期刊格式要求而退回修改

上述两轮审查通过后,主编或副主编会邀请合适的审稿人对论文进行评审。每篇论文至少会邀请两个审稿人来评审,审稿人应是对论文主题较为熟悉的专家,以确保能提供具有参考价值的审稿意见。多数期刊和审稿人都选择采用匿名审稿的方法,即审稿意见不署名,

[Manuscript reference number]
[Manuscript title]

Dear Dr. [Author name],

Thank you for your submission.

Unfortunately it does not appear to be of enough general interest to be suitable for our journal, and accordingly I have decided to reject it without sending it out for review. We receive many more submissions than we can accept. I wish you luck in finding a suitable venue for your work.

Thank you for giving us the opportunity to consider your work.

Yours sincerely,
[Editor name]
Editor
[Journal title]

(a) 主编拒稿

Dear Prof. [Editor name]

Thank you for your message informing us about your decision on our submission without sending it out for review arguing that "it does not appear to be of enough general interest to be suitable for our journal". We fully understand that the final decision lies in your hands. However, we are rather surprised by the argument for not sending it out for review because although there are many studies about …, no such studies exist that focus on …, which is gaining more significance considering what impact global warming (change in permafrost, increase in precipitation, etc.) may have on …. That, we think, would be of significant general interest and was the reason behind our choice of the [Journal title].

Sincerely,
[Author name]

(b) 作者争论

Dear Dr. [Author name],

The decision to reject without review is never one that I make lightly. Typically it has nothing to do with the merits of the scientific arguments in the paper, but concerns more general issues regarding the significance of the results and whether they will be of interest to a broad readership. About half the papers that are submitted to [Journal title] are rejected, often after a lengthy review process. One purpose of rejecting papers upfront is to save time for both the authors and the reviewers. I think it's best to get a quick decision in cases where ultimately the paper is unlikely to be accepted. Inevitably I make some mistakes and a few [Jounral title]-worthy papers are rejected. But there are many geoscience journals and authors can quickly resubmit their papers elsewhere, in many cases to venues that might be more appropriate. In your case, one factor that went into my decision is the fact that you describe computer simulations of … without any specific comparisons to field or laboratory observations. While it is interesting to see how the model behavior changes with respect to the model parameters, the paper is mainly descriptive and lacks any focus on a specific geophysical problem. Thus, it might be better suited to a more technical journal.

Sincerely,
[Editor name]

(c) 主编维持拒稿决定

图 4.5　作者与主编就论文与期刊主题是否相符而争论的案例

作者不知道审稿信息,但审稿人可以看到完整的作者信息。作者和审稿人双向匿名的审稿方式在土木工程领域国际英文期刊较为少见。当然也有少数审稿人愿意在审稿意见后署名。

作为审稿人,首先应有保密意识,不可透露或剽窃任何交由自己审稿的论文内容。评审意见的撰写应包括如下内容:总体概括论文的研究意义、创新点和局限性,英文称为 general comments;然后逐条列出具体审稿意见,包括学术的和写作的意见,英文称为 detailed/technical comments,为方便理解,建议按行号将审稿意见和文中内容相对应。审稿人一般不指出文章的书写问题,而是重点关注文章的条理性、简洁性和正确性。

当主编收到审稿意见后,将据此对论文做出决定。若所有审稿人都双手赞成,论文本身也几乎没有什么需要修改的地方,那么编委直接同意论文发表即可。不过直接接受的情况在学术论文审稿中极其少见。即使所有审稿人都持肯定意见,也会或多或少提供一些修改意见,主编或副主编会给出修改的决定。某些情况下,如果两位审稿人的意见不统一,主编或副主编可将论文再发送给其他审稿人评审,根据多数意见做最后决定;或由自己作为第三审稿人评审,再综合之前的审稿意见做决定。

总体上,主编或副主编对论文的决定可分为五类:接收(accept)、小修(minor revision)、大修(major revision)、拒稿重投(reject & resubmit)、拒稿(reject)。对于前四类审稿意见,作者都应该以积极的心态对论文进行修改完善,多数情况下经过认真修改论文都会被接受。论文的修改方法将在下一节详细介绍。

在审稿过程中,若遇到审稿周期较长的情况,作者也可以主动与负责处理论文的主编或副主编直接联系询问审稿进展,他们一般会乐意并及时回答作者提出的相关问题(图 4.6)。

> Inquire the final decison on the manuscript (manuscript ID).
>
> Dear Dr. [Editor name],
>
> We submitted the revised manuscript "manuscript ID" to [journal name] on [submision date]. We are writing to ask the status of our manuscript and looking forward to your favorable decision.
>
> We appreciate all of your kind help in this manuscript.Thank you very much.
>
> Sincerely yours,
> [Author name]

(a) 催问审稿进度

> Hello Dr. [Author name],
>
> I have to date received only one review for your paper. I have asked 12 different reviewers for a response, and am still waiting for 5 of them to respond. The others all declined to review. This is unusual, but I am doing the best I can.
>
> Best regards,
> [Editor name]

(b) 主编回复示例

图 4.6 作者与主编沟通审稿事宜

4.4 论文修改

当收到邀请小修、大修或改后重投的审稿决定时，作者应仔细研读审稿人的评审意见，确定能否按审稿人提出的意见对论文进行修改。对于多位审稿人同时指出的问题，作者要引起足够的重视，并应遵循他们的意见做出相应的修改，包括研究数据的补充或论点的修改。如果审稿人误解了论文中的有关内容，提的意见并非正确，作者应有理有据且语气客观平和地解释清楚审稿人存在误解的地方。特别是当多位审稿人同时对某一内容都有误解时，说明作者可能对相关内容没有阐述清楚，此时作者一定要慎重，并应对其做出更为明确的说明或解释。

在提交论文修改稿时，期刊通常要求作者提交一份详细的修改说明（图4.7），英文称为

> Dear Prof. Soteris Kalogirou,
>
> Manuscript Number : RENE-D-19-04991
>
> The issues raised by the two reviewers are helpful and constructive. The manuscript has been carefully revised and the following responses to the comments are provided. Note that all the line numbers in the responses are referred to the revised manuscript.
>
> Yours faithfully,
>
> Guihong Liu, Guiling Wang, Zhihong Zhao, Feng Ma
>
> --
>
> **Response to comments from reviewer #1**
>
> **Comment 1.** The abstract can be improved by adding some key figures of their findings
>
> **Response**: A graphic abstract is added in the revised manuscript.
>
> **Comment 2.** The numerical model needs to be validated, this is the major weakness of the paper.
>
> **Response**: We agree! The local scale numerical model for case study is extracted from a city-scale numerical model of the Dezhou geothermal field, China. In Section 3.1 of this revised manuscript, we present the geological and geothermal characteristics of the Dezhou geothermal field, as well as the calibration and preliminary results of the city-scale numerical model of the Dezhou geothermal field. We addressed that the local scale model covering an area of 25 km² is extracted from the numerical model for the Dezhou geothermal field model, to serve as the test model of a heterogeneous geothermal reservoir for case study without losing generality in this generic study. In this way, the computation time can be significantly saved (lines 197-202).
>
> **Comment 3.** Conclusions can be improved
>
> **Response**: The conclusions are rewritten to address the key findings and the underlying mechanisms in Section 6 (lines 392-404).

图 4.7　修改说明示例

a list of responses to the comments。有些期刊还要求提交一份注释版的论文草稿,英文称为 a separate copy of the revised paper in which you have marked the revisions made。当对论文进行修改时,务必注意主编或副主编设定的论文修改截止日期。根据不同的审稿决定,论文修改周期都在 2 周至 2 个月,如果作者认为论文修改时间不够或因某些原因需要延长论文修改时间,均应及时向主编或副主编申请。

4.5 论文发表

经过论文投稿和返修之后,论文如被接受,作者会收到主编或副主编的通知(图 4.8)。论文被接受后,将进入出版阶段。期刊的文字编辑会制作论文的校样稿,作者应仔细审阅校样稿,并认真修改,这是对论文进行修改的最后机会,该过程通常也是以在线修改的模式进行。之后论文将会被正式发表,漫长的投稿工作也就结束了。

[Reference number]
[Paper title]
[Journal title]

Dear Dr. [Author name],

I am pleased to inform you that your paper "[Paper title]" has been accepted for publication in [Journal title].

Your accepted manuscript will now be transferred to our production department and work will begin on creation of the proof. If we need any additional information to create the proof, we will let you know. If not, you will be contacted again in the next few days with a request to approve the proof and to complete a number of online forms that are required for publication.

Effective peer review is essential to uphold the quality and validity of individual articles, but also the overall integrity of the journal. We would like to remind you that for every article considered for publication, there are usually at least 2 reviews required for each review round, and it is critical that scientists wishing to publish their research also be willing to provide reviews to similarly enable the peer review of papers by other authors. Please accept any review assignment within your expertise area(s) we may address to you and undertake reviewing in the manner in which you would hope your own paper to be reviewed.

Thank you very much for expressing your interest in [Journal title].

Sincerely,
[Editor name]
Editor-in-Chief
[Journal title]

图 4.8 论文接受示例

第 5 章　学术会议

学术会议是某一领域的学者和工程技术人员聚集在一起介绍和交流科研成果的活动，也是期刊论文之外的另一重要交流平台。学术会议通常会出版会议论文或摘要集。学术会议按规模可分为大会（symposium 或 conference）和专题研讨会（workshop）。学术会议通常包含少量大会报告（plenary lecture）和分组报告（session presentation），而分组报告可进一步分为口头报告（oral presentation）和海报（poster）。作为青年学者，应充分利用学会会议这一良好的交流平台，与同行分享成果、了解动向、建立友谊，尽早地融入学术领域大家庭。本章主要介绍会议论文撰写与学术会议分组报告时应注意的若干问题。大会报告通常邀请的都是领域内具有一定学术成就的学者，当然有些学术会议也会设置针对青年学者的大会报告专场（plenary session for emerging scientists）。本章介绍的内容同样适用于大会报告。

5.1　会议论文

要想在学术会议上做报告，通常需要提前提交会议摘要或论文。多数学术会议组委会通常会进行两轮评审，即第一轮先提交会议摘要，摘要经评审通过后方可提交会议论文全文（图 5.1）；当会议论文全文通过评审后，学术会议组委会结合作者要求、综合考虑后确定每篇论文的报告形式，比如口头报告或海报。

Key dates

Abstract submission opening:	~~1. Feb 2019~~
Abstract submission deadline:	~~23. Jun 2019~~ Extended to 2. July
Paper submission deadline:	Extended to 10. Nov 2019
Registration opens:	2. Jan 2020
Review notification:	15-24 January
Deadline revised paper submission:	~~2 February~~ Extended to 10 February
Notice on paper acceptance:	~~10. Feb 2020~~ Extended to 21 February
Deadline Earlybird:	8. April 2020
Launch of Mobile App:	Date to be announced

图 5.1　学会会议的关键日期信息示例
(http://www.eurock2020.com/keydates.cfm)

学术会议摘要与期刊论文摘要相类似，但前者的长度限制在 300~500 个单词，通常比期刊论文摘要的 200 个单词长一些。不过，两者的逻辑结构是一致的，只是在学术会议的摘要里能够适当详细地呈现研究背景、方法、结果、结论等内容（图 5.2）。某些情况下，提交学术会议摘要时尚未获得所有的研究结果,此时应在摘要中给出预测的研究结果,并在提交会议论文全文时补充完整。

> **Using inverse modeling to quantify the unknown parameters pertaining to metamorphic fluid flow**
>
> Hide affiliations
>
> Zhao, Z. (*Department of Geological Sciences, Stockholm University, Stockholm, Sweden;*);
> Skelton, A. (*Department of Geological Sciences, Stockholm University, Stockholm, Sweden;*)
>
> Metamorphic reactions can liberate fluid species such as H2O, CO2 and CH4, which are buoyant and migrate upwards through the Earth's crust. To elucidate the role of metamorphism in long term global chemical cycles, it is critical to quantify the flux rate and duration of metamorphic fluid flow and rates of metamorphic reactions and hydrodynamic dispersion in metamorphic fluids. Although time-integrated and time-averaged flux rates of metamorphic fluids and the rates of reactions driven by metamorphic fluid flow have been estimated by reactive transport modeling in the past, most of these estimates rely on simplifications which allow derivation of an analytical solution to some form of the reactive transport equation. These include ignoring the term for hydrodynamic dispersion in the direction of fluid flow, assuming 'grain scale equilibrium' and assuming a 'steady state' or 'quasi-stationary state'. Given that these assumptions are frequently used to quantify metamorphic fluid flow, we consider it useful to verify their usage by 1) developing a general inverse (i.e. back-analysis) modeling framework for quantification of metamorphic fluid flux rates and the rates of fluid-driven metamorphic reactions, and 2) using this framework to validate previous estimates of these parameters which were based on analytical solutions to the reactive transport equation. Our newly developed inverse modeling approach combines numerical solutions of reactive transport equation with the differential evolution method. This general transport model considers advection, hydrodynamic dispersion and fluid-rock reactions under a transient state, which is solved by the Galerkin finite element method. By using global optimization algorithms such as the differential evolution method, those unknown parameters (e.g. flux rates) that cannot be measured in the field or in the laboratory, can be set trial values within certain ranges and continuously adjusted, until the discrepancies between modeling results (e.g. reaction progress) and observed data decrease to a pre-defined tolerance. We assume that the real values for the unknown input parameters are approximated when the discrepancies decrease below this specified tolerance. A MATLAB code was developed to perform this inverse modeling, and was verified by a benchmark test with artificial parameters. To validate the applicability of the 'quasi-stationary state approximation' to metamorphic systems, we re-calculated metamorphic fluid flux rates and flow duration during greenschist facies metamorphism in the SW Scottish Highlands, and compared with the results obtained by assuming a quasi-stationary state. For fluid flow velocities and durations, most of the order-of-magnitude differences are below 30%, which verifies the use of the quasi-stationary state approximation to quantify metamorphic fluid flow at least for the greenschist facies pressure-temperature-composition conditions. For other assumptions such as ignoring the term for hydrodynamic dispersion in the direction of fluid flow and assuming 'grain scale equilibrium', their applicability to metamorphic systems are under investigation.
>
> | | |
> |---|---|
> | **Publication:** | American Geophysical Union, Fall Meeting 2012, abstract id. V23D-2848 |
> | **Pub Date:** | December 2012 |
> | **Bibcode:** | 2012AGUFM.V23D2848Z |
> | **Keywords:** | 1009 GEOCHEMISTRY / Geochemical modeling |

图 5.2　会议摘要示例(EGU2012)

会议论文较一般的期刊论文有严格的页数限制,通常在 4~8 页以内(图 5.3)。会议论文与期刊论文一样都有标题、作者、摘要、关键词、正文(引用、方法、结果、讨论、结论)、致谢、

> **CALL FOR PAPERS**
>
> More than 450 abstracts have been submitted by Authors from all over the world.
>
> The Organising Committee wishes to thank all the Authors for their highly appreciated work.
>
> Notification of acceptance letters have already been sent to all accepted abstract Authors.
>
> The complete list of accepted abstracts is here available for download.
>
> For any further clarification, Authors can contact the organizing secretariat at abstract.iacmag@symposium.it.
>
> Accepted abstract Authors are now invited to submit their paper through the conference submission system.
>
> Papers must be submitted using the conference paper template. Consent to publish to the publisher (Springer) is also required. **Paper length should be between 6 (minimum) to 8 (maximum) pages. Shorter or longer papers will not be accepted.**
>
> Both files (paper template and consent to publish) can be downloaded from the conference submission system.
>
> Deadline for submission: ~~December 15, 2019~~ postponed to December 30, 2019

图 5.3　会议论文长度限制示例(IACMAG 2020 Conference)

参考文献等要素,但没有亮点、附录、补充材料等要素。第1章中介绍的期刊论文撰写原则同样适用于会议论文,但由于篇幅限制,相应的内容都应更精简些。

5.2 口头报告

如同撰写学术论文,学术会议口头报告也应按照研究背景与目的、研究方法、研究结果、讨论与结论的顺序进行。但是学术会议分组口头报告的时长一般为15分钟,因此口头报告无法详细介绍所有的研究细节,也不可能大量引用文献。在组织口头报告的内容时,作者应侧重于研究结果、讨论与结论两方面。作者准备学术会议的口头报告时应该从两方面着手,首先是学者自身的演讲能力,其次是幻灯片(slides)的效果。

(1) 演讲时要做到语速适中,重点突出,勿口齿不清、长篇大论;与观众要有良好的眼神交流,揣摩观众对所演讲内容的理解程度;适时采用身体语言、激光笔等辅助手段强调核心内容,让观众最大限度地理解演讲的核心内容(图5.4);切忌照读幻灯片中的文字。

图5.4 清华大学"科技论文写作与交流"课程邀请专家报告

(2) 幻灯片应根据演讲内容专门设计(图5.5)。在制作幻灯片之前,先要通过会议网站或联系会务组了解投影仪比例,4∶3或是16∶9等。会议室通常较暗,故幻灯片内容与背景的对比度要高,比如白色背景黑色文字;幻灯片版面布置应简明、朴素、整齐、干净,令人赏心悦目;幻灯片中的内容不应太多、太拥挤,切忌堆砌大段文字或大量图表;确保幻灯片中添加的音频、视频等内容可顺利播放,无兼容性等问题;建议主标题至少为20～24号字,副标题至少为16号字,图表名至少为14号字;若某页幻灯片内容信息量大,或内容存在逻辑关系,若一次性展示出全部信息容易混乱观众的思维,则建议添加动画,以便更有条理地进行演示。

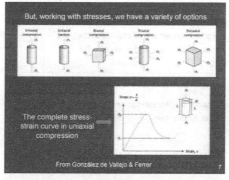

(a) 以图为主

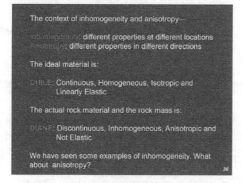

(b) 文字为主

图 5.5 幻灯片示例（来自 John Hudson）

5.3 学术海报

对于大型学术会议,受时间限制,大会不可能为所有收录的会议论文安排口头报告,学术海报就成为另一种会议交流方式。近年来,学术海报这种报告形式在国际会议中变得越来越普遍。在制作学术海报之前,作者先要通过会议网站或联系会务组了解展板尺寸以及所分配的海报编号(图 5.6)。

学术海报应尽量包含会议论文的主要内容,包括引言、方法、结果、讨论、参考文献等,其中科研结果是需要重点介绍的。引言部分,三言两语引出问题;方法部分,简明扼要列出方法要点和特点;结果和讨论部分,概括说明;参考文献部分,列出最重要的几篇即可(图 5.7)。除了在内容上精雕细琢,海报的界面布置其实更需要重视,制作海报要做到清晰醒目、引人入胜。最重要的就是标题,必须简洁,冗长的标题很难让观者驻足,作者姓名略小于标题即可,正文字高一般在 4 mm 左右。海报做好后需根据会议安排,正确编号。标题吸引观者驻足,内容布置就该让观者叹为观止了！精妙绝伦的布置如清风山泉,令人心旷神怡;杂乱无章的布置如一团乱麻,令人纠结不安。简明扼要的语言,形象生动的图示,错落有致,是一股清泉能疏通观者思路;密密麻麻的文字,长篇大论的堆砌,铺满界面,是一潭死水会混乱观者视听。故海报之制作也要恪守"简约"的原则,内容布置要注意留白,能用插图说明的问题,尽量少用大段文字叙述,这样的海报才是成功的海报。

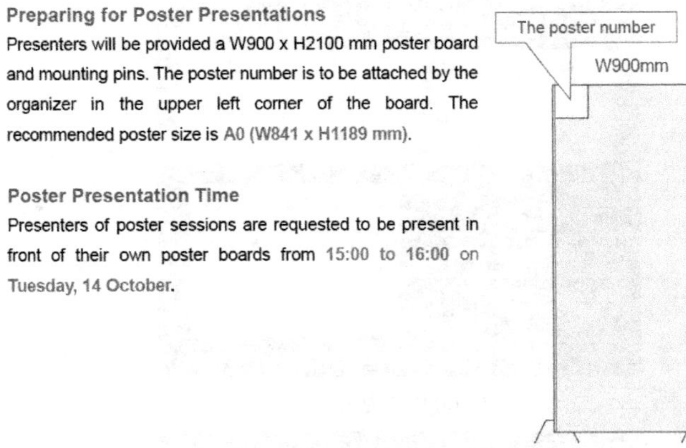

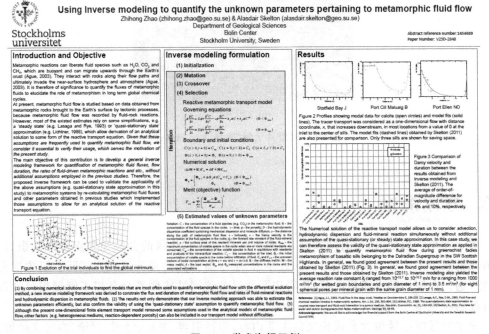

图 5.6 会议网站提供的学术海报制作说明示例（ARMS2014）

图 5.7 学术海报示例

5.4 交流礼仪

　　学术会议参会者可能既是报告人，又是观众。学术会议上展示的内容通常很前沿，故作者对新概念的阐述必须清晰具体。口头报告的观众群体很广泛，很多听众并非本领域研究者，故作者尽量少用非常专业的术语做报告，介绍复杂概念时可以具体一些。问答环节允许报告人与听众有不同见解，并围绕问题展开讨论，但学术讨论不能变成不愉快的争辩，甚至是争吵。

第 6 章 其他写作

除常规学术论文之外,与学术相关的其他写作形式还包括综述论文、短评、简历、项目申请书、推荐信等。这些形式的写作要求与常规的学术论文有较大的不同,本章简要介绍这些形式的写作方法与注意事项。

6.1 综述论文

综述论文并非原创的学术成果,而是对某领域已发表的学术论文进行回顾总结。综述论文不只是对参考文献的简单罗列和概括,而应对该领域在某一时期内取得的成果进行提炼,指出存在的问题,并展望未来潜在的研究方向。综述论文的篇幅通常较长,一般在10~50页。在撰写综述论文之前,应先列好写作提纲,有助于理清写作思路;此外,综述论文一般也会添加目录,方便读者阅读(图 6.1)。

图 6.1 综述论文示例[33]

与常规学术论文相比,综述论文的正文包括引言、讨论、结论,但是一般没有研究方法和结果;讨论部分是其中的重中之重。综述论文读者群体广,故其写作应比常规的科技论文更应注重通俗易懂;最好不用专业术语的缩略词,一旦使用,必须给出详细严谨的解释说明。

6.2 学术短评

学术短评包括书评(book review)、社论(editorial、editor's note)、给编辑的信(letter to editor、discussion and closure)等。学术短评的长度一般在 1000 个单词以内,每种学术短评的侧重各有不同,以下逐一介绍。

(1) 书评:一般针对新近出版的学术专著出版,从第三方的角度介绍书中精华,为作者、出版商、读者提供有价值的反馈或参考意见。书评重在“评”,故书评写作应侧重于对书中重要内容的客观评价,可围绕书的创作目的、创作环境、中心思想等展开写作,也可分析成书的立足点、内容的组织方法(图 6.2),还可知人论世,基于作者所处的时代背景展开分析。

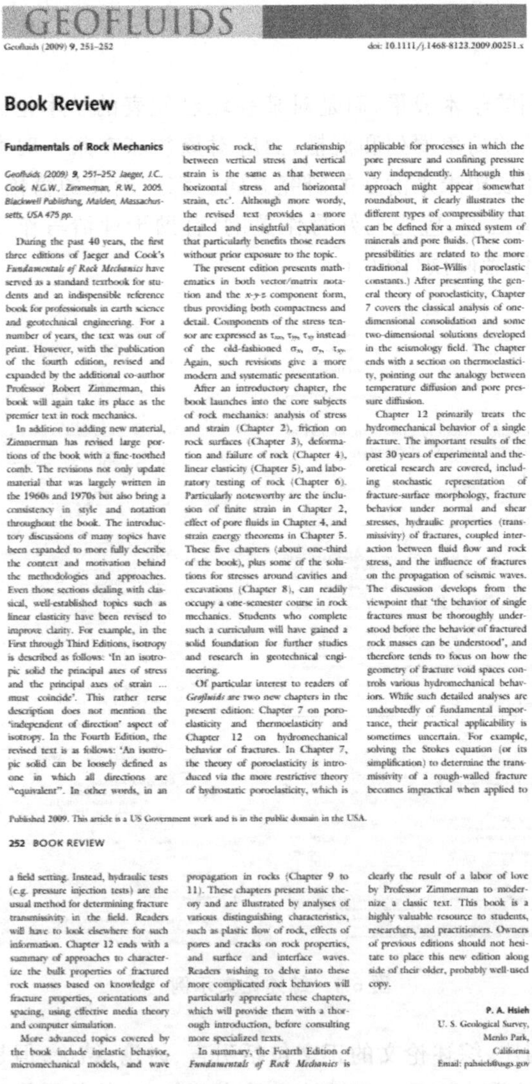

GEOFLUIDS

Geofluids (2009) 9, 251–252 doi: 10.1111/j.1468-8123.2009.00251.x

Book Review

Fundamentals of Rock Mechanics

Geofluids (2009) **9**, *251–252 Jaeger, J.C. Cook, N.G.W., Zimmeman, R.W., 2005. Blackwell Publishing, Malden, Massachussetts, USA 475 pp.*

During the past 40 years, the first three editions of Jaeger and Cook's *Fundamentals of Rock Mechanics* have served as a standard textbook for students and an indispensible reference book for professionals in earth science and geotechnical engineering. For a number of years, the text was out of print. However, with the publication of the fourth edition, revised and expanded by the additional co-author Professor Robert Zimmerman, this book will again take its place as the premier text in rock mechanics.

In addition to adding new material, Zimmerman has revised large portions of the book with a fine-toothed comb. The revisions not only update material that was largely written in the 1960s and 1970s but also bring a consistency in style and notation throughout the book. The introductory discussions of many topics have been expanded to more fully describe the context and motivation behind the methodologies and approaches. Even those sections dealing with classical, well-established topics such as linear elasticity have been revised to improve clarity. For example, in the First through Third Editions, isotropy is described as follows: 'In an isotropic solid the principal axes of stress and the principal axes of strain ... must coincide'. This rather terse description does not mention the 'independent of direction' aspect of isotropy. In the Fourth Edition, the revised text is as follows: 'An isotropic solid can be loosely defined as one in which all directions are "equivalent". In other words, in an isotropic rock, the relationship between vertical stress and vertical strain is the same as that between horizontal stress and horizontal strain, etc'. Although more wordy, the revised text provides a more detailed and insightful explanation that particularly benefits those readers without prior exposure to the topic.

The present edition presents mathematics in both vector/matrix notation and the *x*-*y*-*z* component form, thus providing both compactness and detail. Components of the stress tensor are expressed as τ_{xx}, τ_{yy}, τ_{xy} instead of the old-fashioned σ_x, σ_y, τ_{xy}. Again, such revisions give a more modern and systematic presentation.

After an introductory chapter, the book launches into the core subjects of rock mechanics: analysis of stress and strain (Chapter 2), friction on rock surfaces (Chapter 3), deformation and failure of rock (Chapter 4), linear elasticity (Chapter 5), and laboratory testing of rock (Chapter 6). Particularly noteworthy are the inclusion of finite strain in Chapter 2, effect of pore fluids in Chapter 4, and strain energy theorems in Chapter 5. These five chapters (about one-third of the book), plus some of the solutions for stresses around cavities and excavations (Chapter 8), can readily occupy a one-semester course in rock mechanics. Students who complete such a curriculum will have gained a solid foundation for further studies and research in geotechnical engineering.

Of particular interest to readers of *Geofluids* are two new chapters in the present edition: Chapter 7 on poroelasticity and thermoelasticity and Chapter 12 on hydromechanical behavior of fractures. In Chapter 7, the theory of poroelasticity is introduced via the more restrictive theory of hydrostatic poroelasticity, which is applicable for processes in which the pore pressure and confining pressure vary independently. Although this approach might appear somewhat roundabout, it clearly illustrates the different types of compressibility that can be defined for a mixed system of minerals and pore fluids. (These compressibilities are related to the more traditional Biot–Willis poroelastic constants.) After presenting the general theory of poroelasticity, Chapter 7 covers the classical analysis of one-dimensional consolidation and some two-dimensional solutions developed in the seismology field. The chapter ends with a section on thermoelasticity, pointing out the analogy between temperature diffusion and pore pressure diffusion.

Chapter 12 primarily treats the hydromechanical behavior of a single fracture. The important results of the past 30 years of experimental and theoretical research are covered, including stochastic representation of fracture-surface morphology, fracture behavior under normal and shear stresses, hydraulic properties (transmissivity) of fractures, coupled interaction between fluid flow and rock stress, and the influence of fractures on the propagation of seismic waves. The discussion develops from the viewpoint that 'the behavior of single fractures must be thoroughly understood before the behavior of fractured rock masses can be understood', and therefore tends to focus on how the geometry of fracture void spaces controls various hydromechanical behaviors. While such detailed analyses are undoubtedly of fundamental importance, their practical applicability is sometimes uncertain. For example, solving the Stokes equation (or its simplification) to determine the transmissivity of a rough-walled fracture becomes impractical when applied to

Published 2009. This article is a US Government work and is in the public domain in the USA.

252 BOOK REVIEW

a field setting. Instead, hydraulic tests (e.g. pressure injection tests) are the usual method for determining fracture transmissivity in the field. Readers will have to look elsewhere for such information. Chapter 12 ends with a summary of approaches to characterize the bulk properties of fractured rock masses based on knowledge of fracture properties, orientations and spacing, using effective media theory and computer simulation.

More advanced topics covered by the book include inelastic behavior, micromechanical models, and wave propagation in rocks (Chapter 9 to 11). These chapters present basic theory and are illustrated by analyses of various distinguishing characteristics, such as plastic flow of rock, effects of pores and cracks on rock properties, and surface and interface waves. Readers wishing to delve into these more complicated rock behaviors will particularly appreciate these chapters, which will provide them with a thorough introduction, before consulting more specialized texts.

In summary, the Fourth Edition of *Fundamentals of Rock Mechanics* is clearly the result of a labor of love by Professor Zimmerman to modernize a classic text. This book is a highly valuable resource to students, researchers, and practitioners. Owners of previous editions should not hesitate to place this new edition alongside of their older, probably well-used copy.

P. A. Hsieh
U. S. Geological Survey,
Menlo Park,
California
Email: pahsieh@usgs.gov

图 6.2 书评示例[34]

（2）社论：分为邀请型社论和说服型社论，邀请型社论为期刊内相同领域的科技论文提供评论和背景，同行评审通常会被邀请撰写此类社论；说服型社论用于论证某一特定的观点，包括对该观点和与之对立的观点的赞同与反对（图 6.3）。

Rock Mech Rock Eng (2012) 45:647
DOI 10.1007/s00603-012-0302-6

EDITORIAL

Editorial

Giovanni Barla

Published online: 23 August 2012
© Springer-Verlag 2012

In the modeling of rock mechanics and rock engineering problems, one is faced with the choice of continuum or discrete methods, depending on the problem scale and the geometry of the discontinuous features in rock and in rock masses. Typically, continuum methods are used when no fractures or many fractures are present, whereas discrete methods are more appropriate for describing moderately fractured media containing a limited number of discontinuities.

Continuum modeling by closed-form solutions was used very early in "rock mechanics and the design of structures in rock" (Obert and Duvall 1967), soon to be superseded by numerical methods, mainly the finite element method, FEM (Zienkiewicz and Cheung 1967). With the rock mass being recognized as "made up of a large number of elements… influenced by two basic properties, its immense anisotropy and discontinuity…" (Müller 1974), the attention moved to discontinuum modeling.

With the well-known "Goodman joint element" (Goodman et al. 1968) implemented in FEM codes, discontinuum modeling became possible, although large-scale opening, sliding, and complete detachment was not permitted. It was only with the landmark paper presented in Nancy, France, by Peter Cundall (1971) that the representation of rock masses as "blocky rock systems" was initiated with the discrete element method (DEM).

A remarkable development of DEM took place in the following years (e.g., Jing 2003; Sainsbury et al. 2011). These methods are enjoying, today, a wide application in rock mechanics and rock engineering by the research community and practicing engineers, in 2D and 3D simulations, in static and dynamic conditions. At the same time, efforts are being undertaken to develop alternative formulations for dealing with discontinuum modeling, including fracturing and coupled processes simulations.

The Rock Mechanics and Rock Engineering Journal has edited this special issue with the intent of providing a view of present trends, with a number of recently received papers, where some of the above-mentioned discontinuum modeling aspects are discussed. However, it is not the intent to provide our readers with a comprehensive presentation of the state-of-the-art of this type of modeling.

G. Barla (✉)
Politecnico di Torino, Turin, Italy
e-mail: giovanni.barla@polito.it

References

Cundall PA (1971) A computer model for simulating progressive large scale movements in blocky rock systems. In: Proceedings of the international symposium on rock fracture, Nancy, October 1971. International Society for Rock Mechanics (ISRM), vol 1, paper no. II-8, pp 129–136
Goodman RE, Taylor RL, Brekke TL (1968) A model for the mechanics of jointed rock. J Soil Mech Found Div ASCE 94(SM3):637–659
Jing L (2003) A review of techniques, advances and outstanding issues in numerical modelling for rock mechanics and rock engineering. Int J Rock Mech Min Sci 40(3):283–353
Müller L (1974) Technical parameters of rock and rock masses. In: Rock mechanics, course held in 1972 at the Department of Mechanics of Solids, International Centre for Mechanical Sciences (CISM), Udine, Italy. Springer, New York, pp 16–34
Obert L, Duvall WI (1967) Rock mechanics and the design of structures in rock, 1st edn. Wiley, New York, 650 pp
Sainsbury D, Hart R, Detournay C, Nelson M (2011) Continuum and distinct element numerical modeling in geomechanics. In: Proceedings of the 2nd international FLAC/DEM symposium, Melbourne, Australia, February 2011
Zienkiewicz OC, Cheung YK (1967) The finite element method in structural and continuum mechanics, 1st edn. McGraw-Hill, London, p 272

Springer

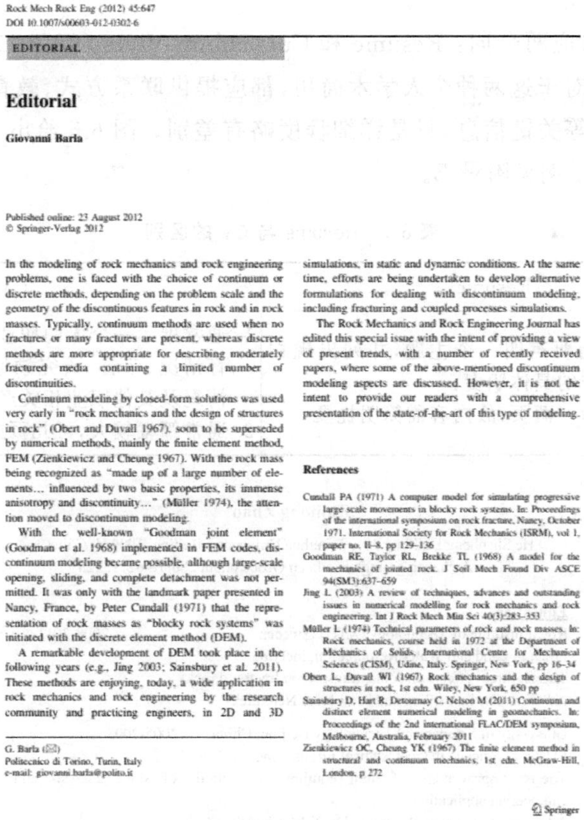

图 6.3 社论示例[35]

（3）给编辑的信：给编辑的信一般是评论已发表文章，以提出有价值的问题为基本准则，被评论文章的作者随后回应评论，其回复与评论一同发表。此类信件通常对近期发表于该期刊的文章进行评论，可作为发表后的同行评议（图 6.4）。

ELSEVIER
Computers & Geosciences 31 (2005) 1305–1307
COMPUTERS & GEOSCIENCES
www.elsevier.com/locate/cageo

Letter to the Editor

Comments on the paper by Pierre D. Glynn, "Modeling Np and Pu transport with a surface complexation model and spatially variant sorption capacities: Implications for reactive transport modeling and performance assessments of nuclear waste disposal sites", Computers & Geosciences 29 (2003) 331–349

In this paper the author uses geochemical transport model to describe the migration of plutonium and neptunium from a deep repository for spent nuclear fuel through a 500 m long flowpath from the repository to the ground surface.

The summary conclusions in the paper are stated as follows.

the water the effect of saturation of the sites will become noticeable.

One of the main themes of the paper is to study how a variation of sorbent concentration along the pathway influences the breakthrough of the trace component. Variations of the concentration of the sorbent are chosen to be stochastic with a log normal distribution. The standard deviations σ are chosen to be 0.5, 1 and 2 in the log scale (I interpret this as meaning 10 log scale). Both the form of the distribution (log-normal) and the values of σ are chosen arbitrarily. Examination of what it means to have a σ of 1, which implies a factor of 10 up and down from the mean, shows that 5% of the path will have a sorbent concentration 100 times less (-2σ) and 5% of the path will have a sorbent concentration 100 times larger (2σ) than the (logarithmic) mean. For

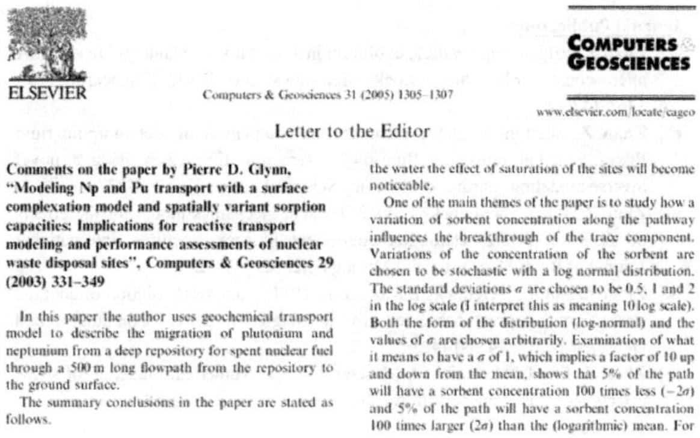

图 6.4 给编辑的信示例[36]

6.3 学术简历与求职信

简历在英文中对应两个词：Resume 和 Curriculum Vitae（CV），二者在内容与结构上有所差别（表 6.1）。对于这两种个人学术简历，都应提供联系方式、教育经历、工作简历、奖励或荣誉、论文发表等关键信息，只是详细程度略有差别。图 6.5 给出了一页 Resume 的示例，更为详细的 CV 示例见附录 B。

表 6.1 Resume 与 CV 的区别

简历	Resume	CV
特点	1. 通常一开始要声明一个主题，并列出现在所从事工作的具体职责； 2. 篇幅一般为一两页，故内容需简明扼要	1. 通常展示一个人的职业生涯：通信信息、受教育情况、荣誉、研究领域、教学情况、发表文章情况及其他职业相关经历； 2. 通常篇幅较大，可几页或更多

Zhihong Zhao

He Shanheng Building 401, Tsinghua University, Beijing, 100084, China
Email: zhzhao@tsinghua.edu.cn Office phone: 86-10-62795223

Education

Royal Institute of Technology (KTH), Sweden   2008-2011
Ph.D. in Land and Water Resource Engineering
Thesis: Stress effects on solute transport in fractured rocks
Advisers: Dr. Lanru Jing and Prof. Ivars Neretnieks

University of Science and Technology Beijing, China   2005-2008
Master of Science in Geotechnical Engineering
Thesis: Improvement of limit equilibrium methods of slope analysis and engineering application
Advisers: Prof. Jin-an Wang and Dr. Y.M. Cheng

University of Science and Technology Beijing, China   2001-2005
Bachelor's Degree in Civil Engineering

Professional positions

Post-doc researcher at Stockholm University, Sweden   2011-2014

Society Memberships

International Society of Rock Mechanics and Engineering 2011-
American Geophysical Union   2011-

Journal Publications

[1] **Zhao Z**. 2013. Gouge particle evolution in a rock fracture undergoing shear: A microscopic DEM study. Rock Mechanics and Rock Engineering. 46: 1461-1479.
[2] **Zhao Z**, Skelton A. 2013. Simultaneous calculation of metamorphic fluid fluxes, reaction rates and fluid-rock interaction timescales using a novel inverse modeling. Earth and Planetary Science Letters. 373: 217-227.
[3] **Zhao Z**, Jing L, Neretnieks I. 2012. Particle mechanics model for the effects of shear on solute retardation coefficient in rock fractures. International Journal of Rock Mechanics and Mining Sciences. 52: 92-102.
[4] **Zhao Z**, Jing L, Neretnieks I, Moreno L. 2011. Analytical solution of coupled stress-flow-transport processes in a single fracture. Computers and Geosciences. 37: 1437-1449.
[5] **Zhao Z**, Jing L, Neretnieks I, Moreno L. 2011. Numerical modeling of stress effects on solute transport in fractured rocks. Computers and Geotechnics. 38: 113-126.

图 6.5 Resume 示例

求职时除提供个人简历外,一般还应附上自述信(图 6.6)。当明确知道收信人时,应在自述信抬头写明收信人,如"Dear Prof. [Family name]";当不知道具体的收信人时,可在自述信抬头写"To Whom It May Concern""Dear Selection Committee""Dear Sir/Madam"。自述信通常只有一页,内容主要包括开头、主体、结尾三部分。开头应开门见山指出所求何

Cover letter
September 10, 2013

Department of Civil and Environmental Engineering
College of Engineering
[University name]

Dear Sir/Madam,

I am writing to apply for an assistant professor of geotechnicalengineering. Now I am a postdoc researcher at [university name], China and studying [research interests].

I am applying for this position motivated by my passion for scientific research in geotechnicalengineering. I studied civil engineering and geotechnical engineering for my bachelor and master Degrees, and then I did my PhD study in engineering geology. Therefore, I accumulated solid fundamental knowledge and broad research visions in the corresponding fields. I am leading a course of applied geology for senior students at [university name], so I also have sufficient pedagogic experiences. By participating in various research projects, I have good communication skills to interact with other colleagues, funding organizations and surrounding society. I am confident to qualify the position of assistant professor in terms of research, teaching and administration, based on my knowledge, education, abilities and experiences.

In this application, I provided the following attachments as you required, from which you can find the details about my application.
(1) CV
(2) Plan for scientific work
(3) Teaching merits

I am excited about the opportunity to join in your department. Please feel free to contact me for more information and interview!

Yours faithfully,
[Applicant name]

图 6.6　求职自述信示例

职；主体部分阐述本人经历和能力与应聘职位要求的吻合度；结尾一般是固定的套话："I believe that my background qualifies me well for the [job name] in your department. I hope to hear from you soon about the possibility of an interview"[综上所述，我相信自己能胜任贵系（职位名称）的工作要求，期待有机会参加面试]。一般不在自述信中直接谈论薪水。对于其他类似在求职成功后再议的事项，建议在面试环节或通过面试之后再与雇主商讨。

6.4 项目申请书

科研工作的开展需要经费，科研经费通常以基金形式，由政府部门、私人基金会或其他机构资助。科研工作者需要申请基金以维持正常的科研运转。科研项目的竞争日益激烈，书写规范是科研申请书的最基本要求，也是保证项目申请成功的基本前提。科研申请书通常应包括以下内容：研究目标、是否在相关机构的资助范围内、研究方法、研究能力与前期工作基础、研究条件、预算等。在正式撰写科研申请书之前，科研工作者应认真研读项目指南，获取项目申请条件，根据自身情况判断是否符合项目要求（图6.7）。

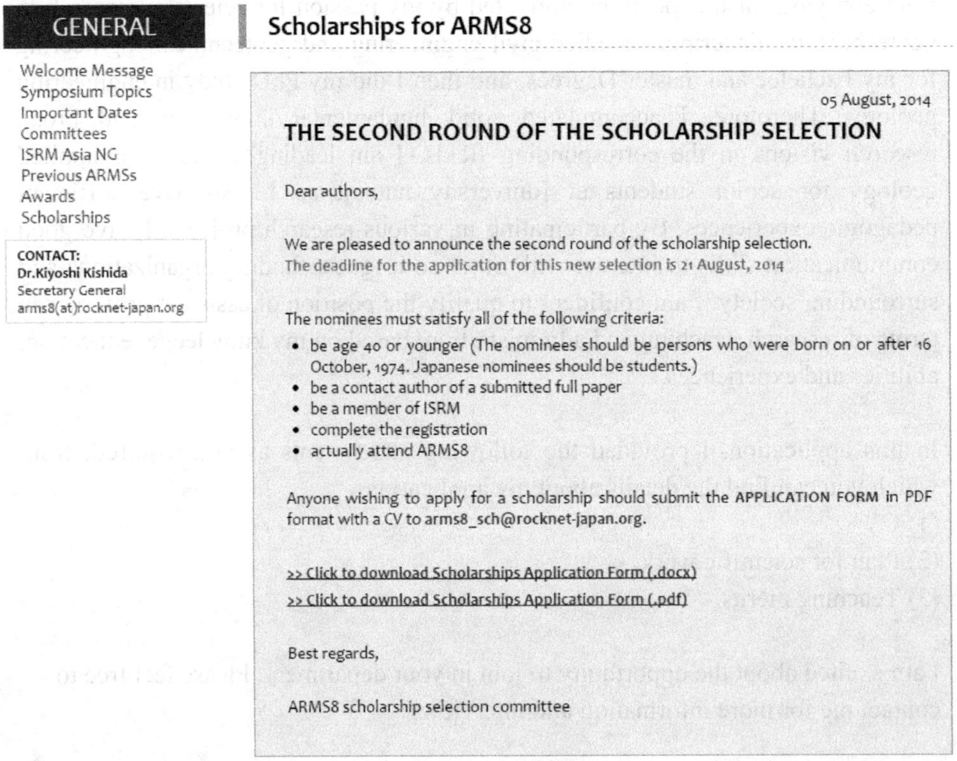

图 6.7　会议奖学金申请条件

如同撰写学术论文时需要参考一些优秀的学术论文一样,撰写科研申请书时若能有一些成功的范例做参考是最好的。一般的项目申请对于申请书的格式都有严格要求,撰写科研申请书应严格按照资助机构提供的模板进行写作(图 6.8)。

> Dear KSRM 2017 Scholarship Selection Committee,
>
> I am writing to apply for the [Name] scholarship, in order to attend the [Conference]. Now I am a PhD candidate at Department of Civil Engineering, Tsinghua University, China, and my main research interests are the coupled thermal-hydro-mechanical-chemical processes in fractured rocks, with application in deep underground engineering.
>
> The [Conference] will be an ideal platform for young scholars of rock mechanics, in terms of exchanging ideas and information and sharing knowledge and experiences. I have attended the first [Conference] in Beijing, China, in 2008, during which I knew my PhD supervisor [Name], and obtained the opportunity to study at the Royal Institute of Technology (KTH), Sweden. Therefore, I do appreciate the series of [Conference] on Rock Mechanics, and would like to contribute to this fantastic conference and look forward to knowing more senior and junior scholars in Jeju Island.
>
> Thank you very much for reviewing my scholarship application.
>
> Sincerely Yours,
> [Name]

图 6.8　会议奖学金申请书示例

6.5　推荐信

无论是帮人写推荐信或是请求别人帮忙写推荐信,都应当思考 3 个问题:被推荐人是否适合这个机会?推荐人是否对被推荐人有足够的了解?应在推荐信中提供哪些信息才能让需求方信服被推荐人的能力?若不能干脆地回答这些问题,推荐人可以婉拒写信的请求。推荐信写作要点如下(图 6.9):

(1) 首段:开门见山地指出,要推荐谁?要推荐他做什么?阐述你和被推荐人的关系(如何认识的)。

(2) 主体部分:全面、客观、公正地描述被推荐人与拟申请的职位或项目相关的优缺点。

(3) 结束语:一般是固定的套话。

(4) 推荐人信息及签名。

此外,推荐信一般应使用官方信纸。

Dear [recipient's name]: or To Whom it May Concern:

It is my pleasure to recommend [Applicant name] for admission to [name of program] at [name of university].I was his/her doctoral supervisor. He/she has an excellent ability to solve difficult physical, mathematical and numerical problems from different disciplines of geomechanics.

[Applicant name]obtained his PhD degree with the thesis titled "[Thesis title]" in 2011. Since then he has been working on [Research subject], with principal applications in geo-energy engineering such as geothermal energy exploitation and underground nuclear waste repositories. He has published [Number] papers and [Number] book chapters. In addition, he has [Number] patents and [Number] software copyrights. These research achievements have been successfully applied to a number of deep subsurface engineering projects, such as [Project titles].

[Applicant name] has a good personality for working with international teams of different backgrounds, expertise and cultures, as demonstrated by his outstanding performance during [project]. At present, he serves as the Editor Board Members of [Journal titile]. He was invited to give plenary talks at [Conferences].

Dr. Zhao was awarded [list of the prizes].

In sum, I consider [Applicant name] an outstanding candidate for [name of opportunity]. I recommend him with enthusiasm.

Sincerely Yours,
[sender's signature]
[sender's name and title]

图 6.9　学术推荐信示例

第 7 章　学术伦理

　　学术论文撰写与发表过程中可能会涉及期刊社/出版社、主编、资助机构、作者等不同机构或个人之间复杂的利益关系。期刊社/出版社负责期刊的出版，为学术论文的发表提供了平台；主编受雇于期刊社/出版社，负责对学术论文的质量进行审核；资助机构为作者提供经费支持；作者将完成的科研工作以论文的形式投稿，经过主编的审核后发表，并将论文版权转移给期刊社/出版社。我们可以看出各群体之间都享有一定的权利，也对其他群体负有相应的责任。为此，有必要明晰上述各方在论文发表中的权利与责任，避免各类学术不当、学术不端等事件的发生。对于学术不端行为，应坚决打击和遏制。

7.1　论文版权

　　版权是复制、出版、出售作品的专属合法权益，版权只保护思想或理念的表现形式，而不保护思想、理念本身。例如：版权不保护你提供的数据，而保护数据的采集过程和得到数据所用的方法。对合作完成的作品，每个参与者共同享有对该作品的版权，并享有相同的权利。

　　作者对自己的论文享有版权，但版权并不是永久的，版权在作者逝世 50 年后自动终止。比如牛顿对其学术成果"牛顿运动定律"的版权在 1777 年就终止了，此后的物理课本就可以"肆无忌惮"地复制这些内容，出版商可以不经牛顿同意直接出版教材，并卖给广大中学生。

　　一般情况下，论文发表后，版权需经由版权所有者，以书面形式转让给期刊社/出版社（图 7.1），此后原作者如果要使用自己的材料，还须从期刊社/出版社处获取权限。这种"古怪"的做法其实是为了赋予期刊社/出版社必要的法律武器。版权转让给期刊社/出版社，期刊社/出版社就可以有力杜绝未经授权而擅用已出版作品的侵权行为，可以有效地维护在期刊上发表作品的全体作者的合法权益。

　　开放获取（open access）是指免费且无限制获取研究成果，例如期刊文章和书籍。对于以开放访问形式发布的学术成果，作者保留版权，但允许期刊社/出版社在特定条件下复制其作品。如果作者使用此类许可在期刊上发表文章，期刊社/出版社要求作者填写诸如协议之类的内容。

　　开放获取有两种形式（表 7.1）：①绿色开放获取（green open access），将期刊文章自存档存入开放信息库；②黄金开放获取（gold open access），在开放获取期刊或提供开放获取选项的期刊上发表，作者需支付出版费用。

CRC Press / Balkema

Consent to Publish & Transfer of Copyright for Contributors to Books

In order to protect the Work against unauthorised use and to authorise dissemination of the Work by means of offprints, legitimate photocopies, microform editions, reprints, translation, document delivery, and secondary information sources such as abstracting and indexing services including data bases, it is necessary for the author(s) to transfer the copyright in a formal written manner.

The <u>Consent</u> ensures that the Publisher has the author's permission to publish the Work.

Title of Contribution

_____

Author(s)

_____

Name of Book / Conference

1. The Author hereby assigns to the Publisher the copyright to the Contribution named above whereby the Publisher shall have the exclusive right to publish the said Contribution in print and in electronic form and translations of it wholly or in part throughout the World during the full term of copyright including renewals and extensions and all subsidiary rights.

2. The Author retains the right to republish the Contribution in any printed collection consisting solely of the Author's own Works without charge and subject only to notifying the Publisher of the intent to do so and ensuring that the publication by the Publisher is properly credited and that the relevant copyright notice is repeated verbatim.

3. The Author guarantees that the Contribution is original, has not been published previously, is not under consideration for publication elsewhere, and that any necessary permission to quote from another source has been obtained. (A copy of any such permission should be sent with this form.)

4. The Author declares that any person named as co-author of the Contribution is aware of the fact and has agreed to being so named.

5. The Author declares that, if the Consent to Publish form has been downloaded from the Publisher's website or sent by e-mail, the form has not been changed in any way without the knowledge of the Publisher.

To be signed by the Author, also on behalf of any co-authors.

Name _____ Date _____

Signature _____

Please return this signed form promptly to the editor of the book / conference. Thank you.

图 7.1 会议论文版权转让协议

表 7.1 开放获取主要形式及其特点

开放获取形式	绿色开放获取	黄金开放获取
权限	某些学术机构或个人网站上免费	发表后对所有人免费
版本	一般为同行评审之前的版本	期刊社/出版社排版的 PDF
费用	免费	收费

关于开放获取期刊负面的评价较多，集中在评审过程相对传统期刊较松、稿源水平整体较低、出版费用较高等方面，建议学者在选择投稿期刊时应慎重选择开放获取期刊。

7.2 发表伦理

学术诚信是指学术机构与学者个人应遵守的客观、诚实、开放、公平、问责与管理的价值观。学术伦理是学术机构之间、学术机构与学者个人、学者个人之间发生关系时应遵循的道德准则，而发表伦理是指在发表论文全过程中涉及的道德准则。发表学术成果的最基本要求是原创性，剽窃、抄袭均是严重的学术不端行为。具体的学术不端行为有：

1）一稿多投与重复发表

一稿多投即将一个稿件同时提交给多个期刊，使用不同语言发表相同内容是常见的一稿多投行为；重复发表是指将要发表的成果与以前发表过的工作，在内容和文章作者上有重复，也就是说，内容大同小异，而且至少包含一个共同作者。

一般来讲，一稿多投和重复发表都是违反学术道德的行为，但在某些特殊情况下也会被允许：以摘要或海报的形式发表，或收录于学会知识库，但前提是获得文章初次发表的期刊的允许。

为避免会议论文发表后再发表期刊论文时发生重复发表的问题，作者首先应获得会议论文发表方的版权许可，而且应在会议论文的基础上增加约 50%以上的新内容，另建议在投稿期刊论文时说明哪些内容已在会议论文中发表过。

如要用另一种语言重新发表已经发表过的研究工作，作者应明确告知两边的期刊编辑，只有编辑们都接受的情况下才可再发表。另外作者在后文中应正确引用已发表的前文，如引用不规范，仍有被撤稿的可能(图 7.2)。

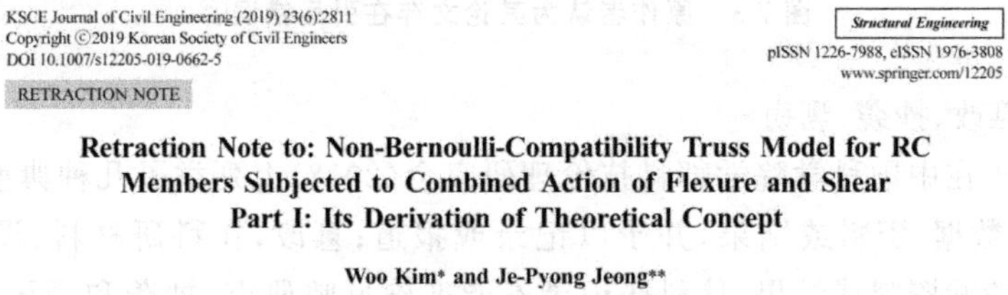
KSCE Journal of Civil Engineering (2019) 23(6):2811
Copyright ©2019 Korean Society of Civil Engineers
DOI 10.1007/s12205-019-0662-5

Structural Engineering
pISSN 1226-7988, eISSN 1976-3808
www.springer.com/12205

RETRACTION NOTE

Retraction Note to: Non-Bernoulli-Compatibility Truss Model for RC Members Subjected to Combined Action of Flexure and Shear Part I: Its Derivation of Theoretical Concept

Woo Kim* and Je-Pyong Jeong**

Received April 29, 2008/Accepted March 16, 2010

The Editor-in-Chief of KSCE Journal of Civil Engineering has decided to retract this article [1] because it has previously been published by Kim *et al.* (2003) [2]. Investigation by the Editorial Board revealed the article [1] is a translation of the entire text from the previously published article [2] without a proper citation. The Editorial Board considers this article to be redundant.

Neither author has responded to any correspondence from the editor about this retraction.

图 7.2　不同语言重复发表案例[37]

2）同行评审不端行为

审稿是学术论文发表前的必经环节，主编通常会邀请对论文研究工作较为熟悉的学者来审稿。多数论文都需要根据审稿意见认真修改后才能正式发表，而且大多数的审稿意见也都是富有建设性的。在土木工程领域，目前的审稿通常以匿名方式进行。对于审稿是否需要匿名的问题，学术界多数人提倡匿名，认为匿名审稿更加公正；有一部分人提倡不匿名，他们认为在审稿意见下署名，是对评论负责的行为。

审稿不端行为包括：审稿人与作者存在利益关系，导致审稿的公平客观性降低；审稿人故意拖延审稿或所提意见明显偏颇；审稿人对作者施压，让其过度引用审稿人的论文；等等。

3）错误引用与不当引用

错误引用是指对所引文献内容评价错误或参考文献标注错误等（图 7.3），而不当引用是指过度引用、漏引等随意性引用行为。除了作者自身存在引用不端行为外，某些期刊为了片面提高影响因子也会对作者施压，让其引用期刊新近发表的论文。

Engineering Fracture Mechanics 232 (2020) 107049

Contents lists available at ScienceDirect

Engineering Fracture Mechanics

ELSEVIER

journal homepage: www.elsevier.com/locate/engfracmech

Letter to the Editor

Letter to the Editor of Engineering Fracture Mechanics

Dear Dr. Ingraffea,

A recent paper in Engineering Fracture Mechanics (EFM) by Niu et al. [1] misrepresented solutions from our EFM paper [2]. The authors of [1] incorrectly state that we consider only one term in the asymptotic expansion, when our contribution is based entirely on the capability to determine the leading two terms for arbitrary loading and arbitrary geometry. Given this capability, the focus of [2] was to determine how far field mixed loading enters into the first term, or possibly the first two terms. For the case of plane stress fracture, the results of [2] demonstrate that the leading term is the pure mode II antisymmetric HRR solution, while the second term is symmetric with an eigenvalue (s_2) close in value to the HRR value (s_1). This means there is no double root in plane stress near mode II, i.e., no mixed mode asymptotic solution. This conclusion was presented in Table 2 and Fig. 15b of [2]. See also [3] for a recent and more complete presentation of this behavior.

As a consequence of their misunderstanding, the authors of [1] made our work appear to be inaccurate, by extracting the wrong values from [2] to include in their Tables 3 and 4 (as well as Tables 1 and 2). For example, in their Table 4 our second term eigenvalue (s_2) for mode II dominant loading was included as though it was our first term eigenvalue (s_1). If they wanted to include correctly our work in their tables, they would have to use s_1 for 360° in Fig. 15a to approximate the mode I values in their Table 3 (which is only available for $m = n = 1, 1.5$ and 3), and s_1 in Fig. 15b for the mode II values in their Table 4. The authors should correct their error, especially since [1] and [2] use the same notation, and Table 2 and Fig. 15 in [2] are very clear about what is s_1 and what is s_2.

Thank you.

References

[1] Niu Z, Li C, Ge R, Hu Z, Hu B. Analysis of plastic stress singularities of cracks and wedges under the plane stress conditions. Eng Fract Mech 2019;208:72–89.
[2] Loghin A, Joseph PF. Asymptotic solutions for mixed mode loading of cracks and wedges in power law hardening materials. Eng Fract Mech 2001;68:1511–34.
[3] Loghin A, Joseph PF. Mixed mode fracture in power law hardening materials for plane stress. J Mech Phys Solids 2020;139:103890.

P.F. Joseph[a], A. Loghin[b]
[a] *Department of Mechanical Engineering, Clemson University, 102 Fluor Daniel EIB, Clemson, SC, United States*
[b] *Simmetrix Inc., Clifton Park, NY 12065, United States*

图 7.3　原作者认为某论文存在引用错误[38]

4）伪造、篡改、抄袭、剽窃

朱邦芬院士在中国科学院学部科技伦理研究会（2018）上列举了几种典型的学术不端行为：伪造，伪造数据、资料或结果，并予以记录或报道；篡改，在科研材料、设备或过程中作假，或者篡改、遗漏资料或结果，使科研记录不能准确反映研究；抄袭和剽窃，窃取他人的思想、方法、成果或文字而未给他人贡献以足够的说明（图 7.4）。

Corrections

Retraction: Zhao, T., W. Liu, L. Zhang, and W. Zhou. 2018. "Cluster Analysis of Risk Factors from Near-Miss and Accident Reports in Tunneling Excavation." *J. Constr. Eng. Manage.* 144 (6): 04018040. https://doi.org/10.1061/(ASCE)CO.1943-7862.0001493.

https://doi.org/10.1061/(ASCE)CO.1943-7862.0001493

This paper has been retracted at the request of the editor. Portions of the paper text are plagiarized from the following published paper:

Raviv, G., B. Fishbain, and A. Shapira. 2017. "Analyzing risk factors in crane-related near-miss and accident reports." *Saf. Sci.* 91 (1): 192–205. https://doi.org/10.1016/j.ssci.2016.08.022.

The editor has concluded that while the two papers are not identical there is enough evidence that the authors of the paper published in the *Journal of Construction Engineering and Management* copied content, concepts, and structure without proper citation and attribution.

图 7.4　自我抄袭案例[39]

5) 署名不当

中国科学院科研道德委员会在《关于在学术论文署名中常见问题或错误的诚信提醒》（2018）中列举署名不当的主要形式：论文署名不完整或者夹带署名；论文署名排序不当；第一作者或通信作者数量过多；冒用作者署名；未充分使用志（致）谢方式表现其他参与科研工作人员的贡献，造成知识产权纠纷和科研道德纠纷；未正确署名所属机构；作者不使用其所属单位的联系方式作为自己的联系方式。

7.3 案例分析

论文发表中学术不端行为、学术不当行为具体表现形式多种多样，本节中简单列举几起学术不端行为、学术不当行为，并建议合理的处理方法。

1) "偷梁换柱"

A博士团队新发表学术论文介绍了一种可用于地下岩土工程示踪试验的新型纳米粒子，该研究还可进一步深入，并有望发表更多的论文。A博士收到了在某大型实验室工作的B博士关于纳米粒子型号与购置渠道的要求，B博士表示希望能重复A博士的试验。相比于A博士较小规模的研究团队，B博士拥有更多资源，A博士认为B博士能用此示踪粒子更快地完成其他研究工作。根据期刊政策，A博士有义务向B博士提供该纳米粒子的相关信息，但A博士向B博士提供了另外一种纳米粒子，却称其为所发表论文中用到的纳米粒子。A博士打算在3个月后，其团队完成后续研究后，再给B博士准确的纳米粒子信息，并承认自己的错误，同时A博士准备为信息错误向B博士道歉。

A博士故意提供给B博士错误的纳米粒子信息，这在学术伦理层面属于学术不端行为，这种故意提供错误信息的行为不利于学术进步。诚然，在学术界存在激烈的学术竞争，因为发表学术成果的最基本要求是原创性，但为了"抢第一"而故意欺骗的行为是不道德的。对于此事件的建议处理方法为，A博士明确向B博士说明自己的顾虑，表示后续研究工作完成后再向B博士提供完整的纳米粒子信息。

2) 评审不端

C博士收到某期刊审稿邀请，对D博士的投稿论文进行评审，C博士与D博士的研究方向相近。在C博士课题组组会上，组内博士后和研究生收到了D博士投稿论文的复印件，C博士打算让其组内成员讨论此投稿论文中未发表的研究结果。

C博士的这一做法违反了期刊审稿的保密要求，在课题组内部就审稿论文进行讨论在学术伦理上是被禁止的。事后C博士意识到自己的行为有失妥当，他回收了所有复印件，并嘱咐其组内成员忽视读过的内容。

3) 署名不当

E博士是某领域德高望重的知名学者，他的博士生F正在开展关于花岗岩岩石力学特性的研究工作。F在该研究上已投入3年多的时间，根据E博士的预期，此研究将产出高水平的研究成果。E博士原计划在该领域重要国际学术会议上报告此项研究的初步成果，同时F也已基于此研究的预期成果，规划了自己的论文。但事与愿违，这项研究并没有按照预期产出任何有价值的成果。E和F都坚信他们的假设无误，但一直无法搞清楚实验究竟出了什么问题。

随着 F 的博士生学习期限日益临近，E 向 F 提议："我有个办法，可以不更换你的论文主题，还能让你在顶级期刊上发文章。"E 将一些材料（试验原始数据和初步的结论草稿）交给 F，这些材料来源于一个已做完的实验，而 F 从未接触过这个实验。E 补充说："我原本计划用几个月的时间写这篇文章，我坚信它可以发表在一个很好的期刊上，但我没有这么多时间，所以写文章一事由你来完成。我去研究我们的试验究竟哪里出了问题，如果我确实找不到问题，你就以这些研究结果为博士课题来毕业答辩。"

如果 F 欣然接受 E 的提议，立即着手撰写文章，而 E 一边试图解决实验的问题，另一边根据自己预期的试验结果，为学术会议撰写摘要，那么二人的行为均属于学术不端。F 把别人的科研成果归于自己，根据"从天而降"的材料写文章，并预期以此获得荣誉，这是典型的学术剽窃。E 先是将不属于 F 的成果作为"礼物"相送，再计划根据预想的实验结果撰写摘要，这是学术腐败和伪造学术成果的行为。

如果 F 一笑拒之："我绝不能在与我毫无关系的科研成果上冠名，如果别人让我解释这些结果，我该如何作答。我会在每个周末重复我的试验，并分析试验结果，直到我搞清楚到底是试验出了差错，还是由于存在某种标定误差，或是我们确实得到了不同于预期的正确结果。除此之外，我不会关注任何其他事情，无论最终结果如何，我都会实事求是地确定我的论文主题。"正直的 F 拒绝了天上落下的"馅饼"，坚持实事求是地开展科研工作，这才是符合学术道德规范的。

4）保密课题

G 被某领域知名学者 H 教授的课题组录取为研究生，并已开始进组开展相关研究工作一年之久。不过她却一直心存疑惑，尽管 H 承担多项基础科研项目，但 G 的研究课题却完全由一家公司资助。来课题组前，G 就向 H 问过这个问题，H 也承诺过公司资助与她进行学术研究并不冲突。但随着工作的进行，G 发现自己整日忙于为公司做实验，根本无暇去研究自己工作中发现的一些有趣的基本问题，或是在其他领域进行探索。尽管在此过程中也学到了很多东西，但她担心自己发表文章有限。由于其他从事公司赞助项目的同学都与相应的公司签署了保密协议，所以无法给她提供有效的参考意见。虽然 H 和该公司的研究者对她的工作非常满意，但她想知道这样的研究状况是否是最适合她的。

H 给 G 安排公司任务是合理的吗？随着 G 继续她的研究，她会在数据收集、数据处理以及成果发表等问题上碰到哪些问题？如果 G 承担的研究工作基于保密要求不能公开发表，对 G 正常攻读博士学位构成了影响，那么 H 的行为属于学术不当。H 应该为研究生合理安排科研课题，在不影响其攻读博士学位的前提下才应介入企业的应用研究。

第8章 文献检索与管理

科技文献的检索和管理贯穿于学术研究全过程,包括科研选题、科研实施以及论文写作或会议交流等环节。科研人员检索得到具有参考价值的文献并加以利用,可以确保学术研究的创新性,避免重复研究。当前,各个领域的论文数量都呈现井喷式增长,科学有效地对检索到的文献进行管理,能够节约时间和经费,助力科研工作的开展。

8.1 文献检索基础知识

1) 文献的定义

在国家标准 GB/T 3792.1—2009《文献著录 第1部分总则》中,文献(document)的定义为:记录有知识的一切载体。知识、记录和载体是构成文献的3个要素。

2) 文献的类型

按照出版形式,常用文献分为以下几类:图书、期刊论文、特种文献、非正式出版物(图 8.1)。

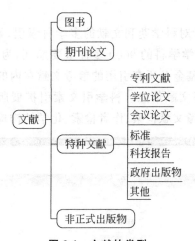

图 8.1 文献的类型

8.2 常用文献资源

国内外文献资源的种类和数量繁多,以清华大学图书馆为例,实体馆藏总量 500 多万册(件);各类数据库 700 多个;电子期刊约 17 万种;电子图书 900 多万册。读者可以根据需要选择。

对于土木工程领域,常用的综合类和专业数据库有:

中文:中国知网(CNKI)、万方数据、维普中文期刊服务平台等。

外文:Web of Science、Engineering Village、ASCE Library、ICE Virtual Library、Scopus、Elsevier ScienceDirect、SpringerLink、ProQuest、EBSCO 等。下面具体介绍前 4 个。

1) Web of Science 简介

Web of Science(http://apps.webofknowledge.com/)是科研人员最常用的信息平台之一,由 Clarivate Analytics(科睿唯安)公司开发,包含 SCI 等多个数据库,信息源于期刊、图书、专利、会议录、网络资源等。

Web of Science 平台包含 Web of Science 核心合集、BIOSIS Previews、中国科学引文数据库(CSCD)、Data Citation Index、Derwent Innovations Index(DII)、Inspec、KCI-Korean Journal Database、MEDLINE、Russian Science Citation Index、SciELO Citation Index 等多个数据库(图 8.2)。

其中,Web of Science 核心合集包含自然科学领域重要的索引数据库——Science Citation Index Expanded(SCI-E,科学引文索引)、Conference Proceedings Citation Index-Science(CPCI-S)、Book Citation Index-Science(BKCI-S),以及社会科学和人文艺术科学领域常用的索引数据库——Social Sciences Citation Index(SSCI)、Arts & Humanities Citation Index(A&HCI)、Conference Proceedings Citation Index-Social Science & Humanities(CPCI-SSH)、Book Citation Index-Social Sciences & Humanities(BKCI-SSH)等。

科学引文索引(SCI)是针对科学期刊文献的多学科索引,始创于 1964 年。其收录了自 1900 年至今 170 多个自然科学学科的 9000 多种重要期刊,为这些期刊的论文编制包括作者、作者单位、摘要、关键词、基金及所有引用的参考文献在内的索引,共收录了 5000 多万篇文献记录及其 10 亿多条参考文献信息。科学引文索引扩展版(SCI-Expanded,SCIE)提供基本检索、高级检索、被引参考文献检索、作者检索、化学结构检索等多种检索方式。

图 8.2 Web of Science 平台

2) Engineering Village 简介

Engineering Village(https://www.engineeringvillage.com/)是爱思唯尔公司出品的信息服务系统,包括 Ei Compendex(美国工程索引,Ei)、Inspec Archive(英国科学文摘,SA)、Knovel 等多个书目文摘数据库。

其中,Ei Compendex 是重要的工程技术文摘数据库,涉及土木工程、机械工程等 190 多个工程技术学科,收录了 1884 年至今的 2700 多万篇文献,文献源自 3000 多种工程类期刊、9 万多种会议以及学位论文、技术标准等。提供 Quick Search、Expert Search、Thesaurus Search 等多种检索方式(图 8.3)。

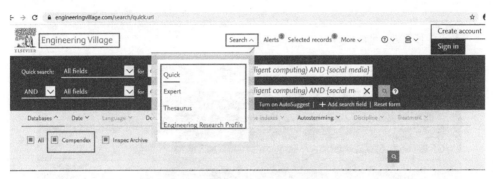

图 8.3　Engineering Village 平台

3) ASCE Library 简介

美国土木工程师学会(American Society of Civil Engineers,ASCE)成立于 1852 年,是历史悠久的国家专业工程师学会。ASCE Library 全文数据库(https://ascelibrary.org/)收录了 39 种 ASCE 期刊、700 多卷会议录,以及图书和标准等,其中 30 种期刊被 SCIE 收录(图 8.4)。作为全文数据库,ASCE 可以让使用者直接浏览感兴趣的期刊或会议录,也可以检索相关文献。对于检索到的结果,科研人员可以导出到文献管理软件,并且可以下载原文。

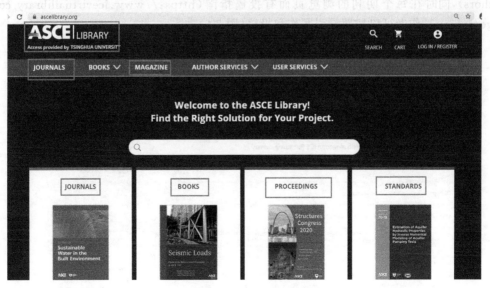

图 8.4　ASCE Library

针对作者的期刊投稿,ASCE 提供专门的投稿指南(https://ascelibrary.org/page/authorservicesjournals),作者可以按照要求写作并投稿。

4) ICE Virtual Library 简介

英国土木工程师协会(Institution of Civil Engineers,ICE)成立于 1818 年,是世界上历史悠久的专业工程机构。ICE Virtual Library(https://www.icevirtuallibrary.com)共有期刊 46 种,其中在版期刊 35 种,2018 年有 20 种 ICE 期刊被 SCIE 收录(图 8.5)。这些期刊涉及土木工程及相关领域,包括 18 种 *ICE Specialist Engineering Journals* 系列期刊以及岩土工程著名期刊 *Geotechnique* 等。

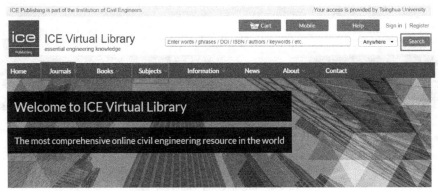

图 8.5　ICE Virtual Library

ICE 作为全文数据库,可以按照期刊进行浏览,也可以按照学科(subjects)领域进行浏览,还可以进行检索。对于检索到的文献,可以下载原文,导出和管理。同时,针对想投稿的作者,ICE 提供了专门的投稿导引栏目(https://www.icevirtuallibrary.com/page/authors),同时在每个期刊的浏览页面有投稿指南(https://www.icevirtuallibrary.com/page/submit),作者可以按照导引写作和投稿(图 8.6)。

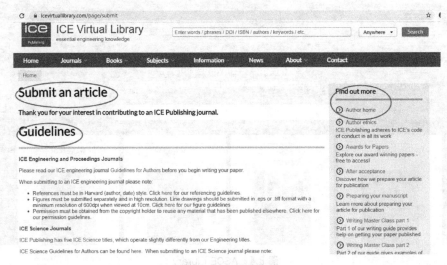

图 8.6　ICE 投稿指南

8.3 文献检索

1)文献检索基本流程

文献检索与管理利用的基本流程包括:分析文献需求;制定、调整和优化检索策略;选择信息源,进行文献检索;分析检索结果,获取原文,对相关文献进行管理和利用;设置文献跟踪等(图 8.7)。其中,检索策略的制定是保证文献检索质量的关键环节。

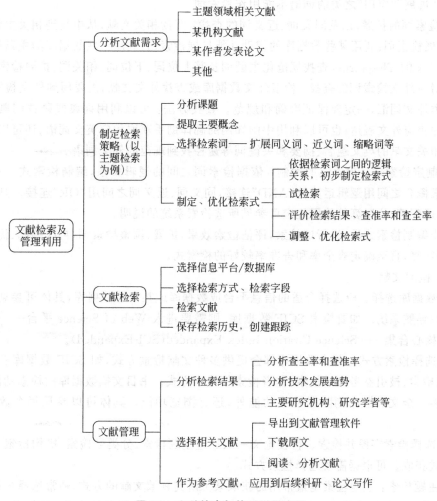

图 8.7 文献检索与管理利用流程

2)文献检索步骤

(1)文献需求分析

文献信息按照记录内容分为研究类文献、数据事实两大类。研究类文献一般利用书目检索工具检索,而数据、事实则利用参考工具检索。

在检索科技文献之前,先要分析文献需求,根据不同需求,选择信息源和检索策略。文献需求一般分为以下几类：课题主题领域的国内外文献、某个研究机构发表的论文和申请的专利、某个学科领域的领军学者发表的论文等。

(2) 制定检索策略

检索策略是保证文献查全、查准的关键环节。检索策略的制定，包括检索词的选取、检索式的制定和优化。针对研究课题，检索策略的制定步骤包括：

① 选择检索词。分析课题，提取主要概念，选择相应的检索词。注意：

a. 检索词的选择原则：要选择具有代表性的、通用的、具有实际检索意义的、科学规范的词语，如具体产品或物质、具体技术或方法、技术参数、性能等。不要用一些泛泛或自创的词语，如"进展""应用"之类的词语不能用作检索词。

b. 检索词的扩展，以及同义词、近义词的查找：阅读相关文献，从中找到相关主题词、关键词，扩展检索词；利用某些数据库的辅助功能扩展检索词；利用一些数据库的主题词表，如 Compendex 的 Thesaurus，查找规范化主题词及其上位词、下位词、相关词，扩展检索词。

c. 对应外文检索词的查找：检索外文数据库或查找外文文献，需要用到外文检索词，对于科技类外文词汇，一定要保证准确和规范。查找方法：可以利用词典或科技词典检索科技词汇的相应外文表达；也可以利用中国知网的翻译助手获取相应英文词语；还可以通过查看中文相关文献的英文题名、摘要和关键词等途径找到规范的外文词语。

② 制定检索式。检索式的编制：依据检索词之间的逻辑关系，编制检索式。一般地，不同检索概念之间用逻辑运算符"AND"连接，同义词、近义词之间用"OR"连接。具体的逻辑算符、通配符、位置算符使用规定可参见所选检索系统的说明。

初步编制检索式后，进行试检索，评估检索效果，扩展、调整检索词，再次编制、优化检索式，循环往复，直至确定查全率和查准率较好的检索式。

(3) 检索文献

① 数据库选择。应选择合适的信息平台或数据库或某个子数据库，具体可参见图书馆的数据库导航系统。如要检索 SCIE 数据库，则需要进入 Web of Science 平台——Web of Science 核心合集——Science Citation Index Expanded(SCI-Expanded)。

② 选择检索方式。每个数据库都会提供多种文献检索方式，如 SCIE 数据库有基本检索、高级检索、被引参考文献检索、作者检索等检索方式。书目文摘数据库的检索功能较强，字段也多。全文数据库则除了检索功能外，还有浏览功能。具体可以参见每个数据库的说明。

③ 选择检索字段并检索。科技论文一般通过主题检索、分类号检索、机构检索、人名检索等方式获取。可根据需要选择检索方式。

a. 主题检索：主题检索是最常用的获取某个研究领域文献的方式，通常选择主题、标题等字段，输入确定好的检索式，其中重要的概念可以选择在标题字段检索。不同数据库对主题字段的定义不同，如 SCIE 中，主题(Topic)字段是指标题、摘要、作者关键词和 KeyWords Plus；Ei Compendex 中，Subject/Title/Abstract(KY)字段则是指在标题、摘要和主题词中检索。具体需要查看各数据库的说明。检索实例见第 15 章。

b. 分类号检索：分类号检索可用于获取某个学科领域的文献，如可用国际专利分类号 IPC、联合专利分类号 CPC 等检索某一技术领域的专利，分类号检索还可与主题检索等其他检索方式联合使用。

c. 机构文献检索：关注本学科领域重要研究机构发表的论文和申请的专利,对于科研创新很有参考价值。

要检索某个机构发表的论文,选择"地址""单位"等表示机构的字段,输入机构名称检索。SCIE 中,可以选择 Address（地址）或 Organizations-Enhanced（机构扩展）字段检索。注意：机构扩展字段,必须采用索引中的正确机构拼写形式,且不能用于院系等二级机构的检索；地址地段,则要注意地址为缩写,缩写形式可以查看地址缩写索引。如要检索一级机构的论文,可以选择地址字段或机构扩展字段,如查找清华大学师生发表的 SCIE 论文,机构扩展的检索式为"TSINGHUA UNIVERSITY",检索得到文献 118 678 篇（2020 年 5 月 30 日检索）。如需获取某个院系发表的论文,则要选择 Address（地址）字段,以清华大学土木系为例,SCIE 中其地址形式有：Tsinghua Univ, Civil Engn Dept; Tsinghua Univ, Dept Civil Engn; Tsinghua Univ, Coll Civil Engn 等。为保证查全,在地址字段输入检索式"(tsinghua univ or tsing hua univ or qinghua univ or qing hua univ) same civil",检索到 SCIE 论文 2062 篇,剔除少量非清华大学论文后得到论文 2050 篇（2020 年 5 月 30 日检索）。Engineering Village 中,机构字段为 Author Affiliation。

如要检索某个单位申请的专利,则选择"专利权人/申请人"字段,输入机构名称或机构代码进行检索。如利用 Derwent Innovations Index 数据库检索清华大学申请的专利,选择"ASSIGNEE NAME & CODE（专利权人）"字段,输入"Univ Tsinghua",检索得到 34 285 个专利（2020 年 5 月 30 日检索）。

d. 个人文献检索：如要了解本学科领域著名学者或学术新秀的研究成果,查找其发表的论文,可选择"Author（作者）"字段,按照数据库要求输入作者名字进行检索。注意：不同数据库,对作者姓名的拼写要求有可能不同。例如：SCIE 中,作者的输入方式是姓氏＋空格＋名字首字母。如中国工程院聂建国院士,在 SCIE 中的作者表达形式为：Nie JG（不区分字母大小写）,但要注意,用"Nie JG"检索到的文献不全是聂建国院士发表的,还须联合作者单位、合作者名字等方式进行限定。

如要查找某个发明人申请的专利,则选择"Inventor（发明人）"字段,输入发明人的名字检索。

e. 期刊论文或会议论文检索：经常查看、跟踪本学科领域重要期刊或会议录的论文,可以了解国内外的最新研究成果,启迪研究思路。

如要获取某个期刊发表的论文,在期刊所在的全文数据库中,可以按照卷期浏览该期刊论文,也可以限制在该期刊内检索；在书目文摘数据库中,选择"Publication Name（出版物名称）""Source Title"等相应字段检索文献。

如土木工程类重要期刊 *GEOTECHNIQUE*,2018 年的 JCR 期刊影响因子为 3.559,在 ENGINEERING、GEOLOGICAL 学科中排序 6/38,Q1 分区。该期刊在 ICE 全文数据库中,可以进入 ICE Virtual Library（https://www.icevirtuallibrary.com/）,按照期刊浏览或检索（图 8.8）。

GEOTECHNIQUE 被 SCIE、Ei Compendex、Scopus 等书目数据库收录。进入 SCIE,选择"Publication Name（出版物名称）"字段,输入"GEOTECHNIQUE",检索得到文献 3943 篇（图 8.9）。

④ 文献管理。对于检索到的文献,需要深入分析,加以管理和利用。

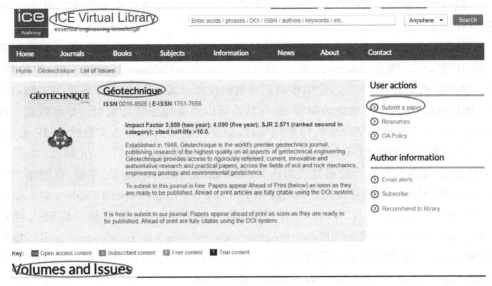

图 8.8　GEOTECHNIQUE

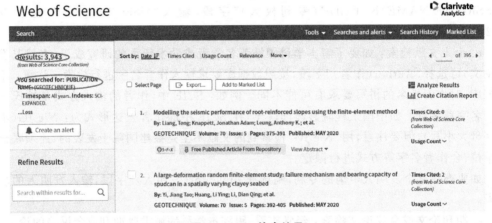

图 8.9　SCIE 检索结果

文献分析：可利用数据库的分析功能，或利用某些软件工具对检索到的文献进行分析，了解国内外研究概况，为科研提供参考。例如：通过出版年分析，可以了解该领域的技术发展趋势和生命周期；按照来源出版物分析，可以了解该领域研究成果发表的主要期刊，为自己的论文投稿期刊提供参考；还可以找到该领域的核心研究单位、重要研究人员等。

选择重要文献，深入阅读。

对于选择的相关文献，导入个人文献管理软件，进行综合管理和利用。

⑤ 设置文献跟踪。及时获取本课题领域的最新文献，有利于开阔视野，促进科研。很多数据库提供个性化服务，如 Web of Science，个人注册账号后，可以保存检索历史、设置跟踪，在对检索式设置跟踪后，系统会自动将最新文献发送到读者设置的邮箱，供读者及时查阅。

8.4 文献管理与利用

科研过程、日常学习和生活中,需要保存和利用的文献信息数量繁多,将其有序、便捷地进行管理并有效利用是非常有必要的,建议选择合适的个人文献管理软件对文献进行管理。

1) 常用文献管理软件

文献管理软件分为两大类:

(1) 单机版:直接安装在个人电脑上。如 NoteExpress(北京爱琴海软件公司)、EndNote 客户端版(Clarivate Analytics 公司)等。

(2) 网络版:基于网络服务器。如 Mendeley(https://www.mendeley.com,Elesvier 公司)、RefWorks(https://refworks.proquest.com,ProQuest 公司)、EndNote Web(Clarivate Analytics 公司)等。

可根据需要选择,其中 NoteExpress、EndNote 客户端版比较常用。

2) 文献管理与利用

(1) 下载安装文献管理软件

有些文献管理软件是免费的,可以注册后下载使用。有的是需要付费的,如 NoteExpress、EndNote 等,如果学校或图书馆已经购买,校内师生可按照说明下载安装后使用。

(2) 创建个人文献数据库

要管理个人文献,首先需创建个人文献数据库。

创建 NoteExpress 数据库:文件→新建数据库→为数据库命名(文件后缀"nel")。

创建 EndNote 数据库:File→New(创建新建数据库 Library,文件后缀"enl")→Groups→Create Group(创建组)、Create Group Set(创建组集合)。

(3) 获取书目数据

获取或添加文献记录,一般有以下几种方式:利用过滤器导入从数据库中检索得到的记录;导入已有的 PDF 全文;在线检索获取书目数据;手工编目等。

(4) 管理个人文献数据库

对数据库中的已有文献记录可以进行管理,如编辑、移动、删除、复制记录,为记录添加笔记、附件(全文、图片等),对记录进行查重、检索等。

(5) 利用个人文献数据库

在论文写作中,编辑部对参考文献有一定的格式要求,作者需要按照要求写作。利用个人文献管理软件,除可从数据库中查找参考文献外,还可以生成特定格式的引文列表,并可以在创作文档中直接插入引文、删除引文、修改引文样式(图 8.10)。另外,EndNote 还有部分内置写作模版可供参考。

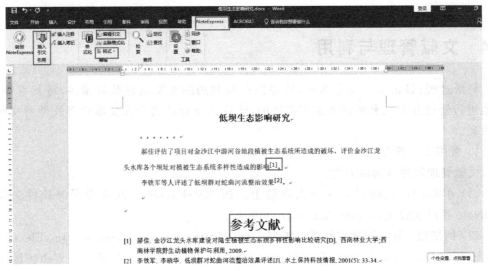

图 8.10　NoteExpress 引文应用

8.5　小结

　　检索本技术领域的科技文献,对其进行分析和管理利用,并贯穿科研过程的每个环节,能够启迪思路,促进科研创新工作的开展。将科研成果以论文形式进行创作,选择合适的投稿期刊,并按照格式要求写作、引用参考文献,论文的发表可使科研成果得到广泛交流和传播,能够促进技术的传承和发展。

实 战 篇

第 9 章　公式编辑指南

MathType 是由美国 Design Science 公司开发的一款专业的数学公式编辑工具,是科研人员专用的工具。MathType 公式编辑器能够帮助科研人员在各种文档中插入复杂的数学公式和符号,与 Office 文档完美结合,显示效果非常好,比 Office 自带的公式编辑器要强大很多。

9.1　公式编辑注意事项

MathType 可以非常便捷快速地输入各种类型的公式,满足科研工作的基本需要,安装成功后可以直接在 Office 工具栏中打开,对于输入各种数学公式十分便捷(图 9.1)。

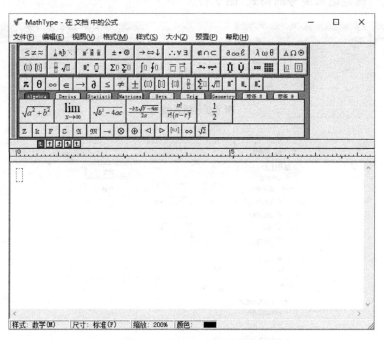

图 9.1　MathType 操作界面

在编辑公式过程中,变量需要用斜体,矢量、张量、矩阵需要用黑斜体,相关格式可以在 MathType 的"样式"中进行选择。变量举例:t;张量举例:**C**。

MathType 的使用注意事项:

(1) 由于版本的不同,会出现旧版本无法打开编辑新版本公式的问题,此时新版编辑后的公式在旧版 MathType 中只是一张图片,在投稿国际期刊中应注意此类问题。

(2) 在 MathType 中输入双层或多层公式,用 Word 编辑时会出现行间距变化的问题,此时推荐使用 Word 自带的公式编辑器。

(3) 在公式输入完成后,有时会对公式中变量或字符进行注释,此时如果使用 MathType 输入字母或字符可能会改变行间距,此时推荐以在 Word 中插入符号的方法解决此问题。

(4) Word 公式编辑器:单击"插入"→"公式"→"插入新公式"可以使用 Word 中自带的公式编辑器,公式编辑器插入后可以输入各类公式满足日常需要(图 9.2)。

图 9.2　Word 公式编辑器界面

9.2　公式的自动编号和引用

步骤①:单击 Word 菜单栏上的"MathType",单击"插入编号"右边的小三角(图 9.3),单击"格式化",确定输入编号的格式(图 9.4)。

图 9.3　插入编号

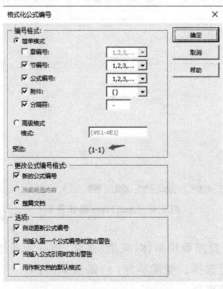

图 9.4　确定输入编号格式

步骤②：单击"MathType"工具栏中的"右编号"，输入公式内容，此时软件开始自动编号，若输入新的公式，则重复步骤②。

$$S\frac{\partial p_f}{\partial t}+\nabla\cdot\left(-\frac{\kappa}{\mu}(\nabla p_f+\rho_f g\ \nabla z)\right)=-\alpha_B\frac{\partial}{\partial t}\varepsilon_{vol} \tag{1-1}$$

$$(\rho C_p)_{eff}\frac{\partial T}{\partial t}+\rho_f C_{p,f}\boldsymbol{u}_f\cdot\nabla T=\nabla\cdot(k_{eff}\ \nabla T)+Q \tag{1-2}$$

步骤③：若要开始新的章节，在上一节最后一个公式后按 Enter 键，单击"MathType"工具栏中的"章 & 节"，选择"插入下一个分节符"(图 9.5)，此时软件会自动开始新的章节编

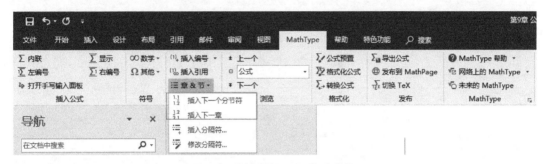

图 9.5　选择"插入下一个分节符"

号，重复步骤②输入公式即可。

$$S\frac{\partial p_f}{\partial t}+\nabla\cdot\left(-\frac{\kappa}{\mu}(\nabla p_f+\rho_f g\ \nabla z)\right)=-\alpha_B\frac{\partial}{\partial t}\varepsilon_{vol} \tag{1-1}$$

$$(\rho C_p)_{eff}\frac{\partial T}{\partial t}+\rho_f C_{p,f}\boldsymbol{u}_f\cdot\nabla T=\nabla\cdot(k_{eff}\ \nabla T)+Q \tag{1-2}$$

$$-\nabla\cdot\sigma=\rho_{ave}\boldsymbol{g}=(\rho_f\varphi+(1-\varphi)\rho_s)\boldsymbol{g} \tag{2-1}$$

步骤④：对公式进行编号后，若要在正文中引用公式，单击"插入引用"，得到"公式参考此处"，此时双击需要引用的公式编号就可以在此处得到对公式的引用(图 9.6)。

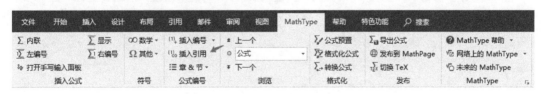

图 9.6　引用公式

$$S\frac{\partial p_f}{\partial t}+\nabla\cdot\left(-\frac{\kappa}{\mu}(\nabla p_f+\rho_f g\ \nabla z)\right)=-\alpha_B\frac{\partial}{\partial t}\varepsilon_{vol} \tag{1-1}$$

$$(\rho C_p)_{eff}\frac{\partial T}{\partial t}+\rho_f C_{p,f}\boldsymbol{u}_f\cdot\nabla T=\nabla\cdot(k_{eff}\ \nabla T)+Q \tag{1-2}$$

$$-\nabla\cdot\sigma=\rho_{ave}\boldsymbol{g}=(\rho_f\varphi+(1-\varphi)\rho_s)\boldsymbol{g} \tag{2-1}$$

引用公式参考此处

$$S\frac{\partial p_\mathrm{f}}{\partial t}+\nabla\cdot\left(-\frac{\kappa}{\mu}(\nabla p_\mathrm{f}+\rho_\mathrm{f} g\ \nabla z)\right)=-\alpha_\mathrm{B}\frac{\partial}{\partial t}\varepsilon_\mathrm{vol} \qquad (1\text{-}1)$$

$$(\rho C_\mathrm{p})_\mathrm{eff}\frac{\partial T}{\partial t}+\rho_\mathrm{f} C_{\mathrm{p},\mathrm{f}}\boldsymbol{u}_\mathrm{f}\cdot\nabla T=\nabla\cdot(k_\mathrm{eff}\ \nabla T)+Q \qquad (1\text{-}2)$$

$$-\nabla\cdot\sigma=\rho_\mathrm{ave}\boldsymbol{g}=(\rho_\mathrm{f}\varphi+(1-\varphi)\rho_\mathrm{s})\boldsymbol{g} \qquad (2\text{-}1)$$

引用公式(2-1)。

第 10 章 图片制作指南

有位论文审稿人在自己的博文中写道："我审稿时看稿件的顺序是题目、摘要、图表、前言、参考文献和正文。"可见论文中图片的质量是非常重要的,处理一张图可能会花费大量的时间,但图片质量的好坏在一定程度上决定了论文能否被录用。本章将介绍科技论文中图片的处理方法及常用的绘图软件,希望能为读者写出更完美的论文提供一些参考。

10.1 数据图

Origin 是美国 OriginLab 公司推出的数据分析和科技作图软件,也是广泛流行和国际科学出版界公认的标准作图工具,功能强大且操作简便,既适合一般的作图需求,也能够满足复杂的数据分析和图形处理,是科学研究和工程研究人员必备的软件之一。Origin 软件界面如图 10.1 所示。

图 10.1　Origin 软件界面

下面将以几种常见的数据图(折线图、散点图、柱状图、Y 误差图、极坐标图、双 Y 轴图和局部放大图)为例进行详细的操作步骤介绍。

1) 折线图

步骤①：单击菜单栏"文件"→"导入"导入数据到 Origin 工作表中，双击"C(Y)"列，在弹出的"列属性"对话框中(图 10.2)，在"选项"→"绘图设定"的下拉列表中，选择"X"，单击"确定"，并按照同样的方法，分别将 E、G、I、K、M、O 列的"列属性"设置为"X"。

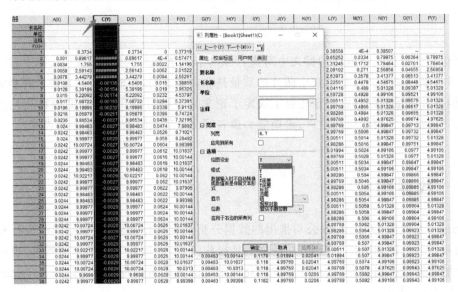

图 10.2 "列属性"对话框

步骤②：全选数据表(图 10.3)，并单击左下角"折线图"图标，软件会自动跳转到 Gragh 窗口(图 10.4)。

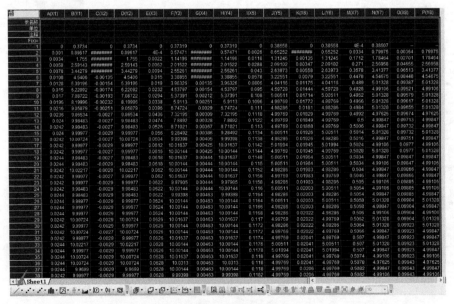

图 10.3 数据表

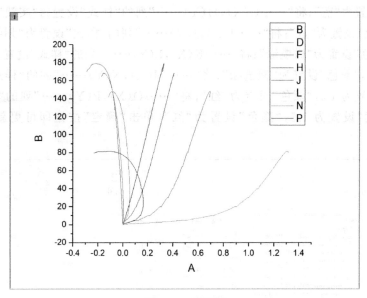

图 10.4 折线图

步骤③：双击图中任意一条曲线，在弹出的"绘图细节-绘图属性"对话框（图 10.5）中将"组"选项卡中的"编辑模式"设置为"独立"，单击"应用"按钮。

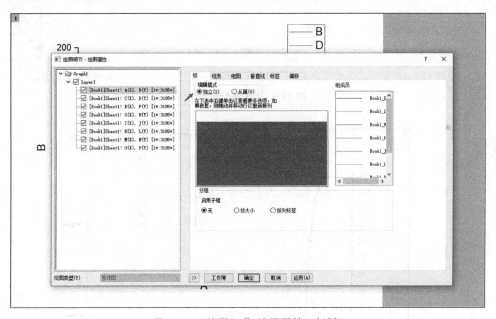

图 10.5 "绘图细节-绘图属性"对话框

步骤④：在"绘图细节-绘图属性"对话框中，单击并进入"线条"选项卡（图 10.6），将"……A(X)，B(Y)……"列的"样式"设置为"划线"，"宽度"设置为 1.5，"颜色"设置为"黑"。按照同样的方法，将"……C(X)，D(Y)……"列的"样式"设置为"实线"，"宽度"设置为 1.5，"颜色"设置为"黑"；将"……E(X)，F(Y)……"列的"样式"设置为"划线"，"宽度"设置为

1.5,"颜色"设置为"蓝";将"……G(X),H(Y)……"列的"样式"设置为"实线","宽度"设置为"1.5","颜色"设置为"蓝";将"……I(X),J(Y)……"列的"样式"设置为"划线","宽度"设置为 1.5,"颜色"设置为"橄榄绿";将"……K(X),L(Y)……"列的"样式"设置为"实线","宽度"设置为 1.5,"颜色"设置为"橄榄绿";将"……M(X),N(Y)……"列的"样式"设置为"划线","宽度"设置为 1.5,"颜色"设置为"红";将"……O(X),P(Y)……"列的"样式"设置为"实线","宽度"设置为 1.5,"颜色"设置为"红",单击"确定"按钮即可更新 Graph 窗口（图 10.7）。

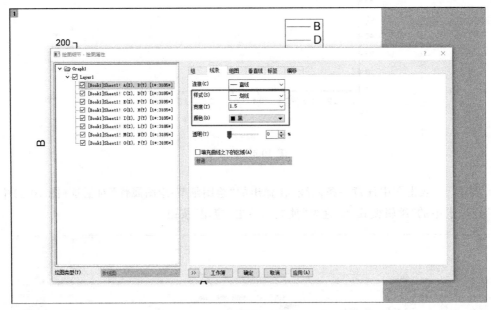

图 10.6 "线条"选项卡

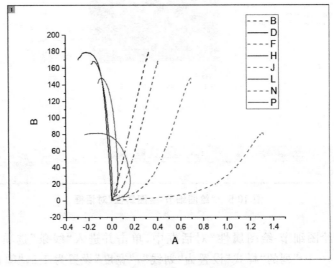

图 10.7 更新后的折线图

步骤⑤：双击"图例"文字，在弹出的"文本对象-Legend"对话框（图 10.8）中，删除多余图例，只保留"\l(1)……"和"\l(2)……"，再修改文字部分内容，然后单击并进入"边框"选项卡（图 10.9），将"边框"设置为"无"，并单击"确定"按钮。

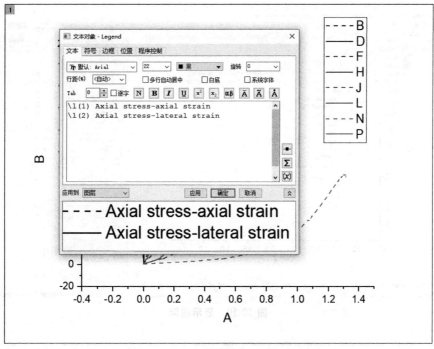

图 10.8 "文本对象-Legend"对话框

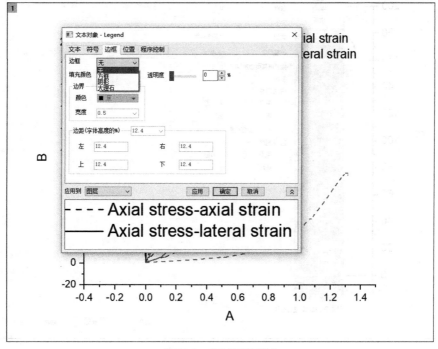

图 10.9 "边框"选项卡

步骤⑥：依次单击"菜单栏"→"查看"→"显示"→"框架"，即可显示框架（图 10.10），将图例文字大小设置为 18，并拖动图例到合适位置，双击"Y 轴"，在弹出的"Y 坐标轴"对话框中，将"刻度"的起始设置为 0（图 10.11）。

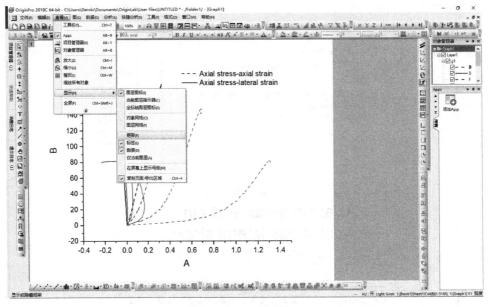

图 10.10　显示框架

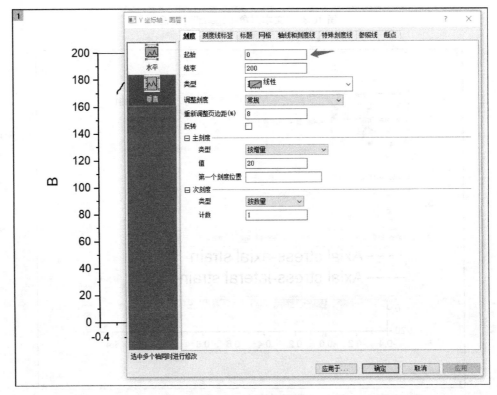

图 10.11　设置刻度

步骤⑦：单击左边"工具栏"中的"直线工具"按钮 ╱，按住键盘上的 Shift 键，在"0.0"处画一条垂直线，双击这条直线，在弹出的"对象属性"对话框中将"宽度"设置为 1.5，单击"确定"按钮。

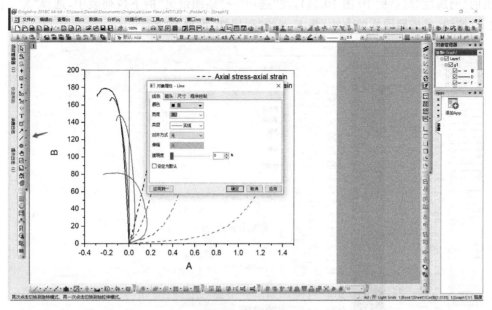

图 10.12　设置宽度

步骤⑧：双击"Y 轴"标题"B"，将文字内容改为"Axial stress（MPa）"，删除"X 轴"标题"A"，单击左边"工具栏"中的"文本工具"按钮 T，给"X 轴"添加两段图 10.13 中的文字，将文字大小设置为 20，并将文本拖拽到合适的位置，按照同样的方法，可在图 10.13 中添加更多的文字注释，即可完成绘图。

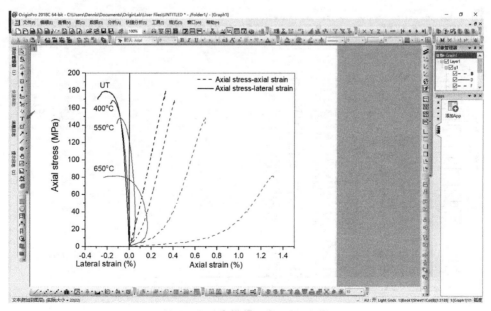

图 10.13　添加文字注释

2) 散点图

步骤①：单击菜单栏"文件"→"导入"导入数据到 Origin 工作表中，双击"C(Y)"列，在弹出的"列属性"对话框中（图 10.14），在"选项"→"绘图设定"的下拉列表中，选择"X"，单击"确定"按钮，并按照同样的方法，将 E 列的"列属性"设置为"X"。

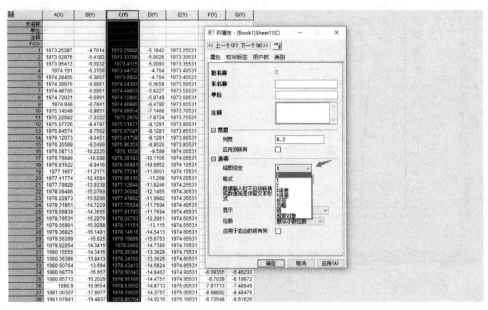

图 10.14 "列属性"对话框

步骤②：选中数据表的 A～C 列（图 10.15），并单击左下角的"散点图"图标 ∴，软件会自动跳转到 Graph 窗口（图 10.16）。

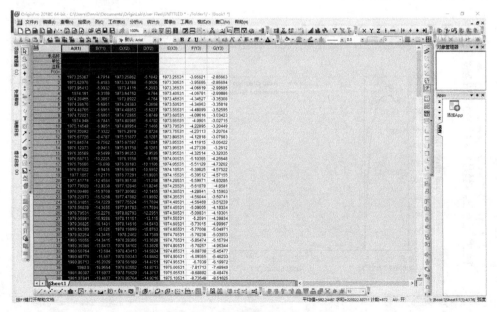

图 10.15 选中数据表的 A～C 列

第 10 章　图片制作指南　**91**

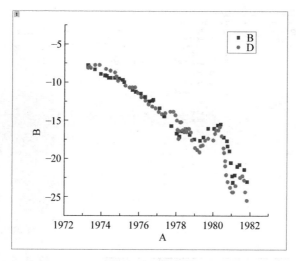

图 10.16　散点图

步骤③：双击图中任意一个散点，在弹出的"绘图细节-绘图属性"对话框（图 10.17）中单击并进入"符号"选项卡，将"大小"设置为 5，"内部"设置为"虚心"，单击"确定"按钮。

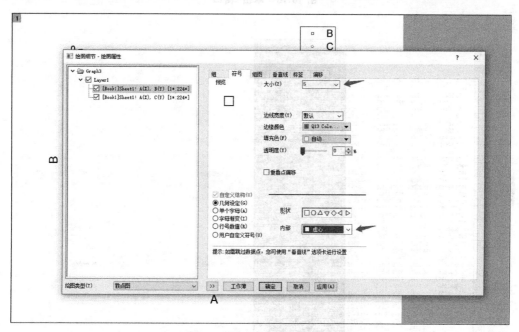

图 10.17　"绘图细节-绘图属性"对话框

步骤④：单击菜单栏中的"窗口"（图 10.18），选择并返回"Book1"，选中数据表的 E～G 列（图 10.19）。再单击菜单栏中的"窗口"，选择并返回"Graph1"。

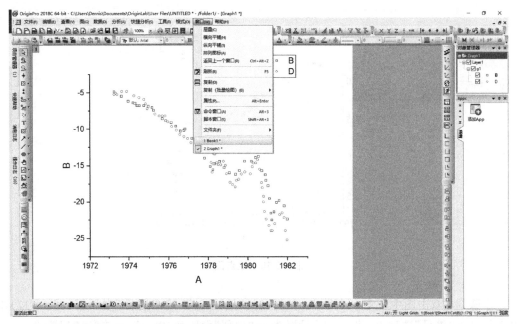

图 10.18 单击“窗口”

	A(X1)	B(Y1)	C(X2)	D(Y2)	E(X3)	F(Y3)	G(Y3)
长名称							
单位							
注释							
F(x)=							
1	1973.25387	-4.7914	1973.25862	-5.1842	1973.25531	-3.95621	-2.85663
2	1973.62976	-5.4183	1973.33788	-5.0026	1973.30531	-3.95665	-2.85694
3	1973.95413	-6.0932	1973.4115	-5.2093	1973.35531	-4.06619	-2.99685
4	1974.191	-6.3159	1973.64752	-4.764	1973.40531	-4.06791	-2.99886
5	1974.28455	-6.3867	1973.8922	-4.764	1973.45531	-4.34527	-3.35308
6	1974.38676	-6.6961	1974.24383	-5.3658	1973.50531	-4.34963	-3.35818
7	1974.48765	-6.6961	1974.48853	-5.6227	1973.55531	-4.48099	-3.52595
8	1974.72021	-6.6961	1974.72865	-5.8749	1973.60531	-4.09616	-3.03423
9	1974.948	-6.7841	1974.80985	-6.4792	1973.65531	-4.0901	-3.02715
10	1975.14548	-6.9851	1974.89954	-7.1466	1973.70531	-4.22895	-3.20449
11	1975.22082	-7.3322	1975.2979	-7.8724	1973.75531	-4.23113	-3.20704
12	1975.67726	-8.4787	1975.51877	-8.1281	1973.80531	-4.12918	-3.07683
13	1975.84574	-8.7562	1975.67597	-8.1281	1973.85531	-4.11915	-3.06422
14	1976.12073	-9.0451	1975.81758	-8.1281	1973.90531	-4.27339	-3.2612
15	1976.35589	-9.5499	1975.96353	-8.9526	1973.95531	-4.32514	-3.32035
16	1976.58713	-10.2225	1976.1558	-9.599	1974.00531	-5.10365	-4.26648
17	1976.76686	-10.098	1976.39183	-10.1106	1974.05531	-5.51129	-4.73202
18	1976.81622	-9.9416	1976.56981	-10.6952	1974.10531	-5.39825	-4.57522
19	1977.1657	-11.2171	1976.77291	-11.0001	1974.15531	-5.39512	-4.57155
20	1977.41774	-12.4584	1976.95138	-11.268	1974.20531	-5.59971	-4.83285
21	1977.79928	-13.9338	1977.12846	-11.8246	1974.25531	-5.61979	-4.8581
22	1978.09486	-15.0769	1977.30082	-12.1455	1974.30531	-4.28941	-3.15903
23	1978.22973	-15.5268	1977.47882	-11.9982	1974.35531	-4.56044	-3.50741
24	1978.31851	-14.7229	1977.75524	-11.7694	1974.40531	-4.56469	-3.51239
25	1978.56839	-14.3655	1977.91793	-11.7694	1974.45531	-5.09005	-4.18334
26	1978.79531	-15.2276	1978.02793	-12.2951	1974.50531	-5.09831	-4.19301
27	1979.05991	-15.9288	1978.11151	-13.115	1974.55531	-5.2591	-4.39834
28	1979.36825	-16.1401	1978.14616	-14.5413	1974.60531	-5.73015	-4.99967
29	1979.56399	-15.625	1978.19899	-15.8763	1974.65531	-5.77008	-5.04971
30	1979.82254	-14.3415	1978.2462	-14.7389	1974.70531	-5.76238	-5.03933
31	1980.15555	-14.3415	1978.28386	-13.3628	1974.75531	-5.85474	-5.15794
32	1980.36386	-13.8413	1978.34102	-13.3628	1974.80531	-5.70257	-4.96344
33	1980.50764	-13.694	1978.43413	-14.5824	1974.85531	-6.08708	-5.45477
34	1980.68779	-15.657	1978.50343	-14.8462	1974.90531	-6.09355	-5.46233
35	1980.85713	-16.2029	1978.56189	-14.4751	1974.95531	-6.7039	-6.19972
36	1980.9	-16.9554	1978.63552	-14.8713	1975.00531	-7.81713	-7.48949
37	1981.00307	-17.6977	1978.76629	-14.3757	1975.05531	-8.68892	-8.48476
38	1981.07841	-19.4837	1978.86764	-14.9216	1975.10531	-8.73548	-8.51625

图 10.19 选中数据表 E～G 列

步骤⑤：依次单击菜单栏上的“图”→“当前图层内添加”→“折线图”，双击图中任意一条曲线(图 10.20)，在弹出的“绘图细节-绘图属性”对话框中，将“组”选项卡中的“编辑模式”设置为“独立”，单击并进入“线条”选项卡(图 10.21)，将“……E(X)，F(Y)……”的“样式”设置为“划线”，“宽度”设置为 1.5，“颜色”设置为“黑”，将“……E(X)，G(Y)……”的“样式”设置为"划线","宽度"设置为 1.5,"颜色"设置为"红",单击"确定"按钮。

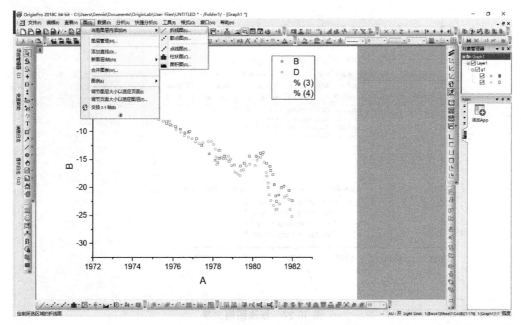

图 10.20 双击曲线

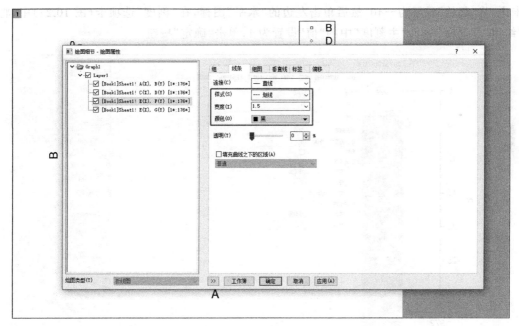

图 10.21 "线条"选项卡

步骤⑥:依次单击"菜单栏"→"查看"→"显示"→"框架",即可显示框架,双击图例文字,将字体大小设置为 18,然后按图 10.22 修改图例文字,单击"确定"按钮,并拖拽图例到合适位置。

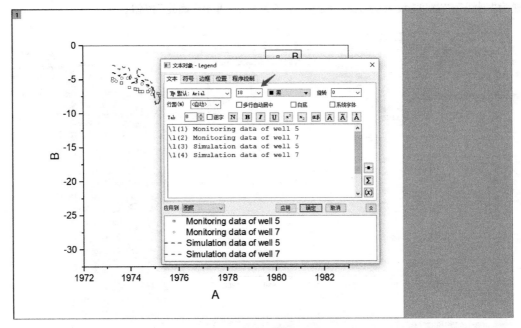

图 10.22　修改图例文字

步骤⑦：双击"Y 轴"，在弹出的"Y 坐标轴-图层 1"对话框（图 10.23）中，在"刻度"选项卡中，将"起始"设置为 -40，然后单击左边的"水平"图标，在"刻度"选项卡（图 10.24）中，将"结束"设置为 1982，"主刻度"中的"值"设置为 1，单击"确定"按钮。

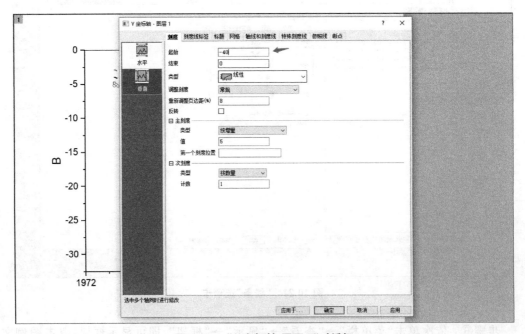

图 10.23　"Y 坐标轴-图层 1"对话框

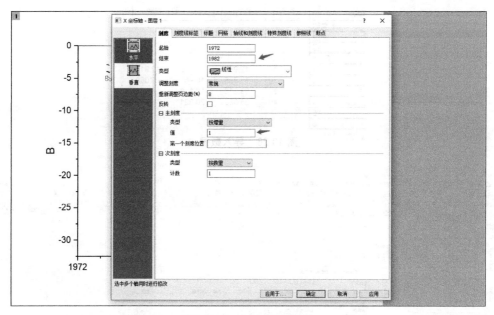

图 10.24 "刻度"选项卡

步骤⑧：分别双击"X 轴标题文字"和"Y 轴标题文字"，按图 10.25 修改文字内容，即可完成绘图。

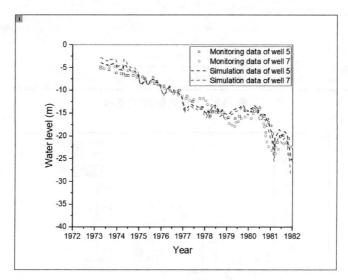

图 10.25 修改文字

3) 柱状图

步骤①：单击菜单栏"文件"→"导入"导入数据到 Origin 工作表中（图 10.26）。

步骤②：单击左下角的"柱状图"图标 ，在弹出的"图表绘制"对话框（图 10.27）中，将"A"列勾选为"X"，其余列勾选为"Y"，单击"确定"按钮，软件自动跳转到 Graph 窗口（图 10.28）。

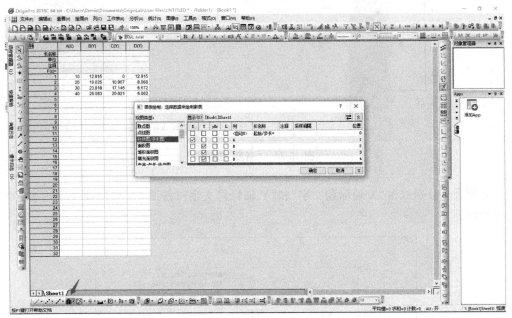

图 10.26 导入数据

图 10.27 "图表绘制"对话框

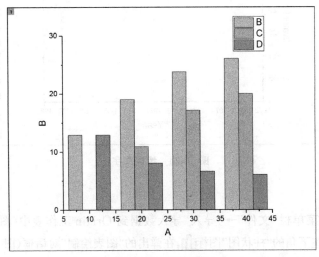

图 10.28 柱状图

步骤③：双击任意柱状图，在弹出的"绘图细节-绘图属性"对话框中，单击箭头所指的按钮，在弹出的"增量编辑器"（图10.29）中，将颜色设置为以下3种，单击"确定"按钮。

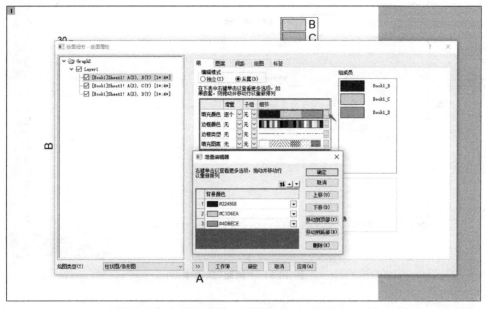

图 10.29 "增量编辑器"

步骤④：单击并进入"图案"选项卡（图10.30），将"宽度"设置为0.2。单击并进入"间距"选项卡（图10.31），将"柱状/条形间距（%）"设置为40，单击"确定"按钮。

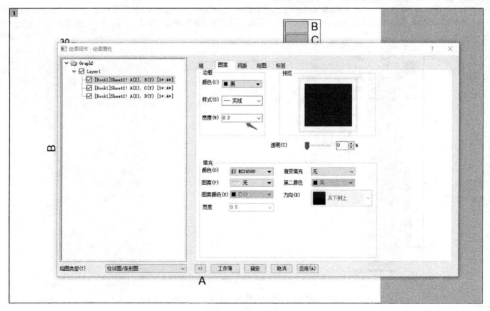

图 10.30 "图案"选项卡

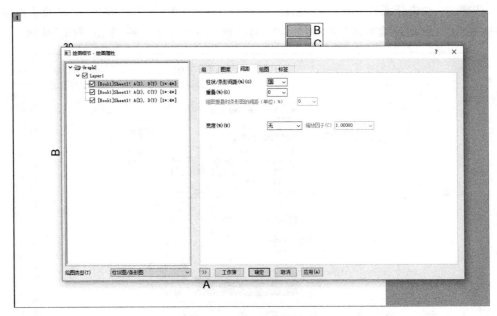

图 10.31 "间距"选项卡

步骤⑤：双击"Y 轴"，在弹出的"Y 坐标轴-图层 1"对话框（图 10.32）中，将刻度的"结束"范围设置为 35，"主刻度"的"值"设置为 5。单击"水平"按钮，在"刻度"选项卡（图 10.33）中将"起始"设置为 4，"结束"设置为 46，"主刻度"的"值"设置为 20，单击"确定"按钮。

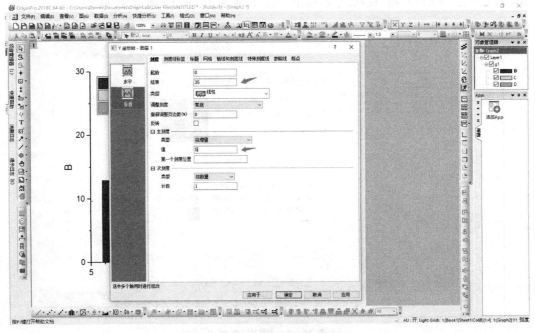

图 10.32 "Y 坐标轴-图层 1"对话框

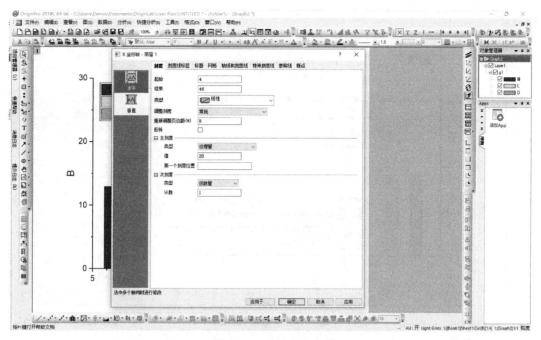

图 10.33 "刻度"选项卡

步骤⑥：双击"图例"文字,在弹出的"文本对象-Legend"对话框中,修改图例文字(图 10.34),并将文本大小设置为 18。单击并进入"边框"选项卡(图 10.35),将"边框"设置为"无",单击"确定"按钮,将图例拖拽到合适的位置。

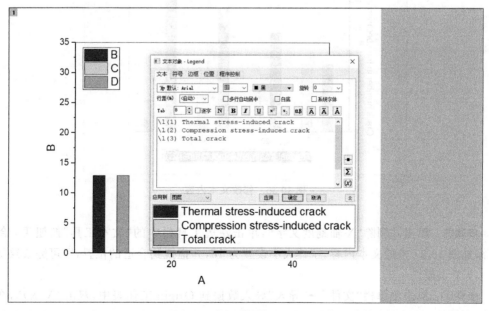

图 10.34 修改图例文字

步骤⑦：分别双击"X 轴标题文字"和"Y 轴标题文字",并按图 10.36 修改文本内容。

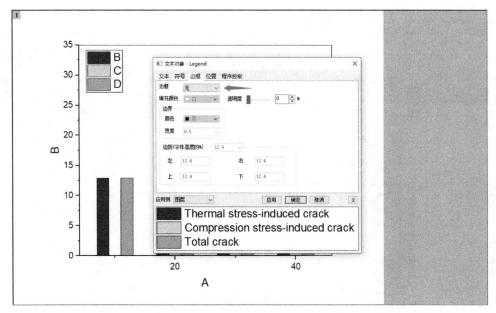

图 10.35 "边框"选项卡

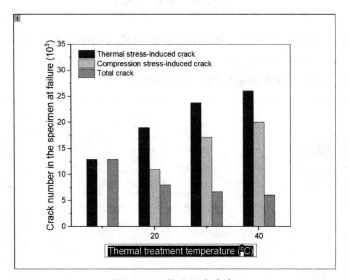

图 10.36 修改文本内容

步骤⑧：单击并删除"X 轴刻度文字"，单击左边工具栏中的"文本工具"按钮 T，给"X 轴"添加图 10.37 中的文本内容，字体大小设置为 18，并拖拽到合适的位置，即可完成绘图。

4) Y 误差图

步骤①：单击菜单栏"文件"→"导入"导入数据到 Origin 工作表中，双击"A(X)"，在弹出的"列属性"对话框中将"选项"中的"格式"设置为"文本"(图 10.38)，双击"C(Y)"，在弹出

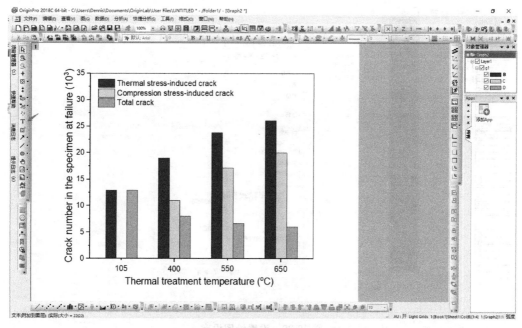

图 10.37 给"X 轴"添加文本内容

的"列属性"对话框中将"选项"中的"绘图设定"设置为"Y 误差"(图 10.39),按照同样的方法,将"E(Y)"列设置为"Y 误差",单击"确定"按钮。

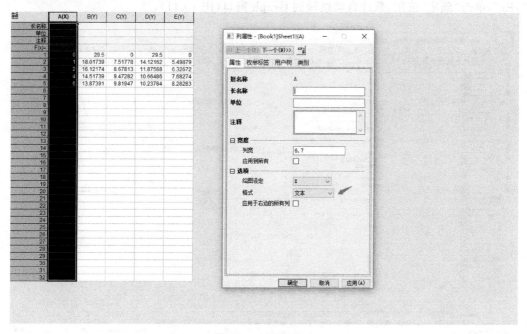

图 10.38 设置格式

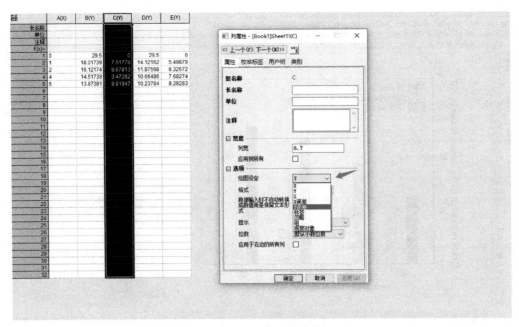

图 10.39　设置绘图设定

步骤②：单击左下角的"点线图"图标，在弹出"图形绘制：选择数据来绘制新图"对话框（图 10.40）中，将"A 列"设置为"X"，"B 列"和"D 列"设置为"Y"，"C 列"和"E 列"设置为"yEr"，单击"确定"按钮，软件自动跳转到 Graph 窗口（图 10.41）。

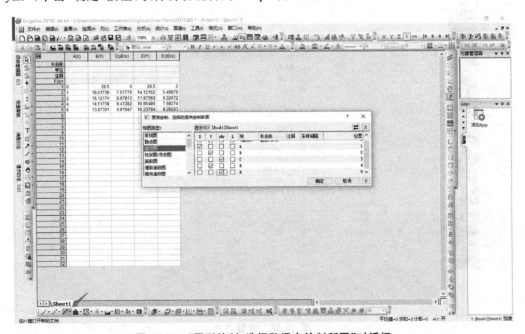

图 10.40　"图形绘制：选择数据来绘制新图"对话框

第 10 章　图片制作指南　103

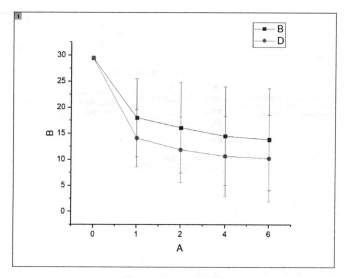

图 10.41　Y 误差图

步骤③：双击图中任意一条曲线，在弹出的"绘图细节-绘图属性"对话框中，在"组"选项卡（图 10.42）中，将"编辑模式"设置为"独立"；然后单击并进入"线条"选项卡（图 10.43），将"……A(X)，D(Y)……"的"样式"设置为"划线"；然后单击并进入"符号"选项卡（图 10.44），将"……A(X)，B(Y)……"的符号设置为"空心矩形"，将"……A(X)，D(Y)……"的符号设置为"空心圆"，单击"确定"按钮。

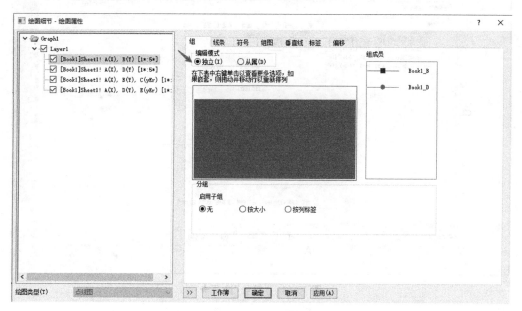

图 10.42　"组"选项卡

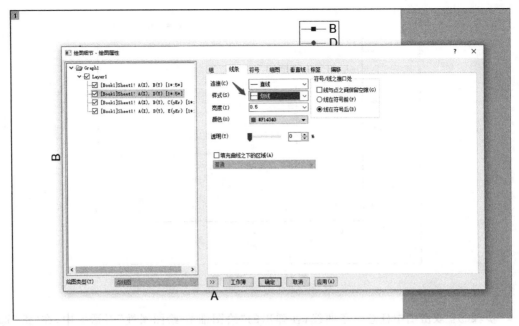

图 10.43 "线条"选项卡

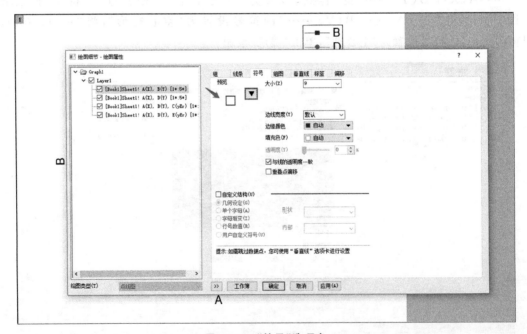

图 10.44 "符号"选项卡

步骤④：单击"查看"→"显示"→"框架"，双击"图例文字"，在弹出的"文本对象-Legend"对话框中，将"字体大小"设置为 18，并按图 10.45 修改图例文字。

步骤⑤：双击"Y 轴"，在弹出的"Y 坐标轴"对话框中，在"刻度"选项卡（图 10.46）中，将"起始"设置为 0，"结束"设置为 35，单击"确定"按钮。

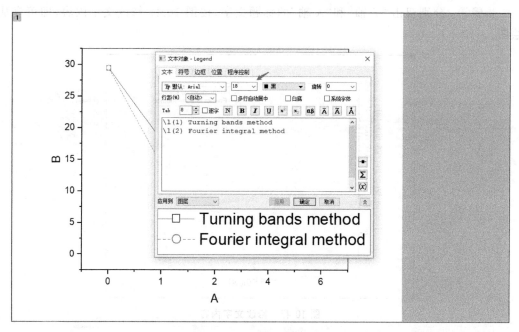

图 10.45 修改图例文字

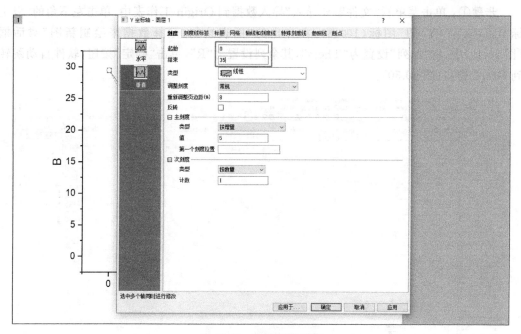

图 10.46 "刻度"选项卡

步骤⑥：分别双击"Y轴"和"X轴"的标题文字，按图10.47修改文字内容，即可完成绘图。

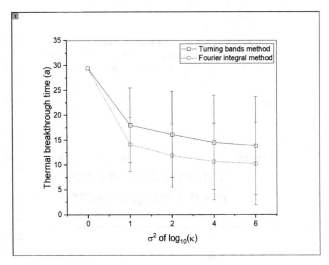

图10.47　修改文字内容

5）极坐标图

步骤①：单击菜单栏"文件"→"导入"导入数据到Origin工作表中，单击左下角的"极坐标theta(X) r(Y)图"图标（10.48），在弹出的"图表绘制：选择数据来绘制新图"对话框（图10.49）中，将"A列"设置为"Theta"，其余列设置为"R"，单击"确定"按钮，软件自动跳转到Graph窗口（图10.50）。

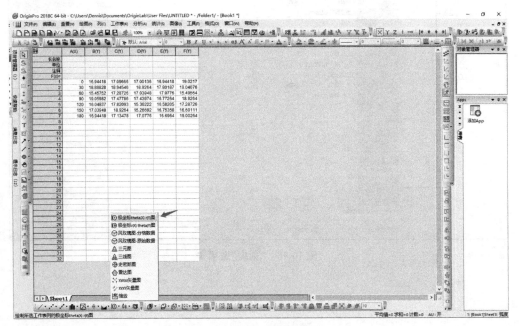

图10.48　导入数据

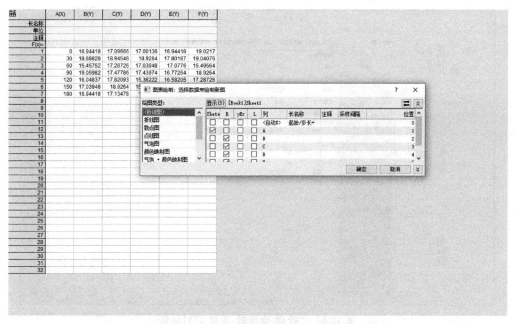

图 10.49 "图表绘制:选择数据来绘制新图"对话框

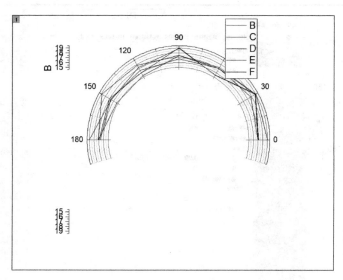

图 10.50 极坐标图

步骤②:双击"径向坐标轴",在弹出的"径向 坐标轴-图层 1"对话框(图 10.51)中,在"刻度"选项卡中去掉"中心在零点"后面的"√",将"起始"设置为 14,"结束"设置为 20。单击"角度"图标,在弹出的"角度 坐标轴-图层 1"对话框(图 10.52)中,将"起始"设置为 0,"结束"设置为 180。

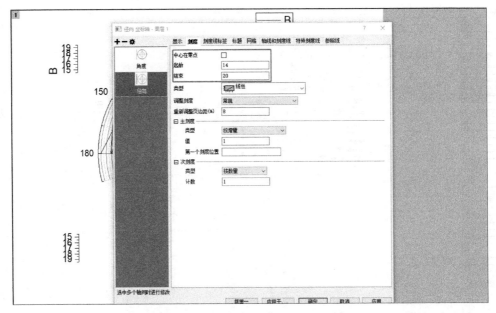

图 10.51 "径向 坐标轴-图层 1"对话框

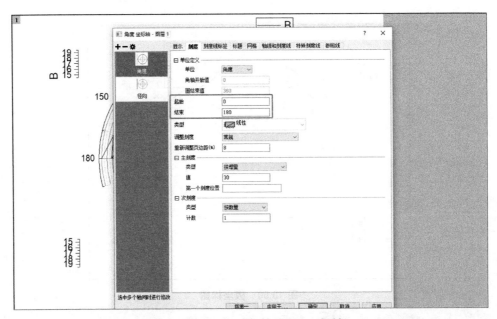

图 10.52 "角度 坐标轴-图层 1"对话框

步骤③：单击并进入"显示"选项卡(图 10.53)，单击"径向-外 2"图标，并去掉"显示"后面的"√"，单击"确定"按钮。

步骤④：双击图中任意一条曲线，在弹出的"绘图细节-绘图属性"对话框(图 10.54)中，将"……A(theta),B(r)……"的"宽度"设置为 5，"颜色"设置为"浅灰"；将"……A(theta),C(r)……"的"宽度"设置为 1.5，"颜色"设置为"灰"；将"……A(theta),D(r)……"的"宽度"

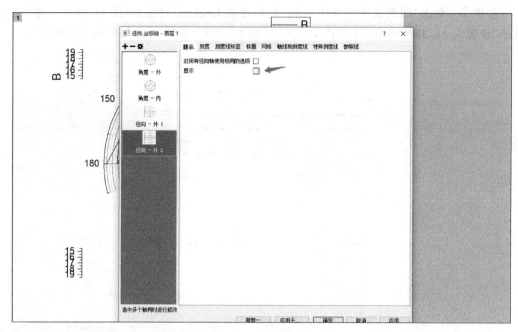

图 10.53 "显示"选项卡

设置为 1,"颜色"设置为"深灰色";将"……A(theta),E(r)……"的"宽度"设置为 1,"颜色"设置为"红";将"……A(theta),F(r)……"的"宽度"设置为 1,"颜色"设置为"蓝",单击"确定"按钮。

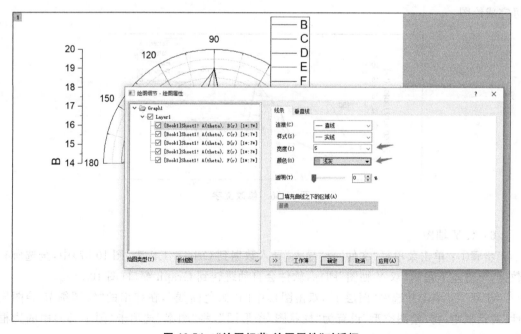

图 10.54 "绘图细节-绘图属性"对话框

步骤⑤：双击"图例文字"，在弹出的"文本对象-Legend"对话框（图10.55）中，将"文字大小"设置为18，并按图10.55修改文字内容，单击"确定"按钮，并拖拽图例到合适位置。

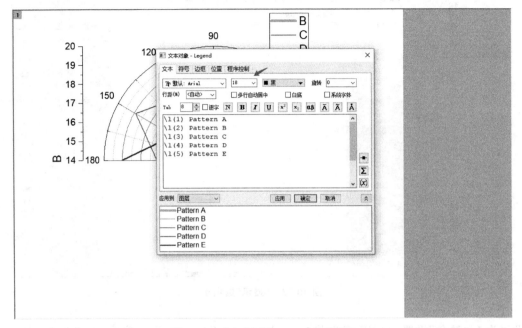

图 10.55　"文本对象-Legend"对话框

步骤⑥：双击"Y 轴"标题文字，按图10.56修改文字内容，并拖拽标题到合适位置，即可完成绘图。

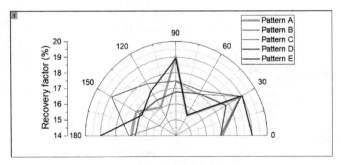

图 10.56　修改文字

6）双 Y 轴图

步骤①：单击菜单栏"文件"→"导入"导入数据到 Origin 工作表（图10.57）中，全选所有数据，单击左下角的"双 Y 轴图"图标，软件会自动跳转到 Graph 窗口（图10.58）。

步骤②：单击并选中"图层 1"，双击图形中的"黑色曲线"，在弹出的"绘图细节-绘图属性"对话框中，将"绘图类型"设置为"柱状图/条形图"，将"图案"选项卡（图10.59）中的边框"宽度"设置为1，填充"颜色"设置为"无"，单击"确定"按钮。

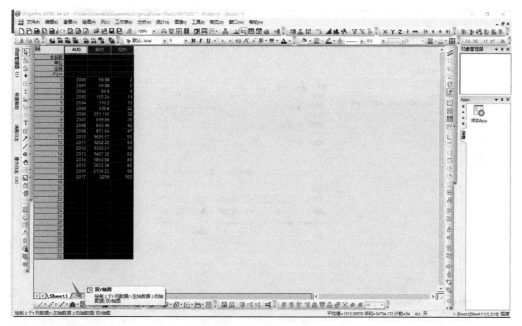

图 10.57 导入数据

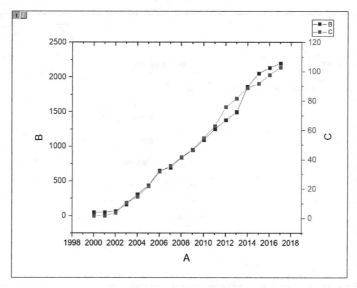

图 10.58 双 Y 轴图

步骤③：单击并选中"图层 2"，双击图形中的"红色曲线"，在弹出的"绘图细节-绘图属性"对话框中，将"绘图类型"设置为"散点图"，在"符号"选项卡（图 10.60）中，将"符号颜色"设置为"红色"，单击"确定"按钮。

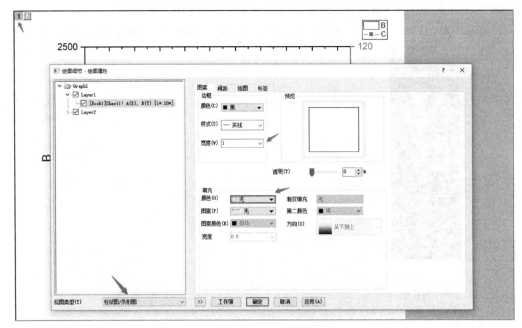

图 10.59 "图案"选项卡

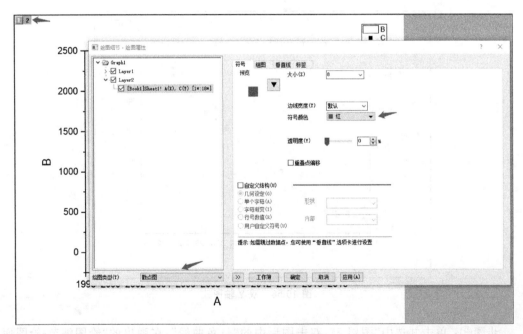

图 10.60 "符号"选项卡

步骤④：双击左边的"Y 轴"，在弹出的"Y 坐标轴-图层 1"对话框中，单击并进入"轴线和刻度线"选项卡（图 10.61），将"主刻度"和"次刻度"的"样式"均设置为"朝内"；然后单击"下轴"图标，并将"主刻度"和"次刻度"的"样式"设置为"朝内"，单击"确定"按钮，按照同样的方法设置右边的"Y 轴"的"样式"。

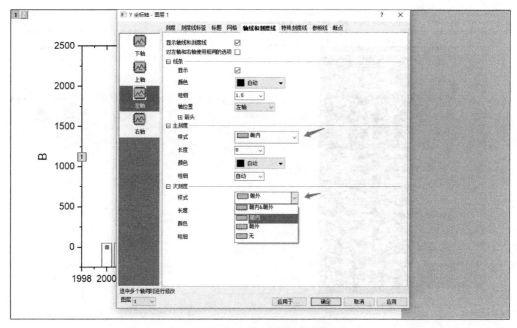

图 10.61 "轴线和刻度线"选项卡

步骤⑤：分别双击"Y 轴"和"X 轴"的标题文字，按照图 10.62 修改文字内容，单击并删除图例，即可完成绘图。

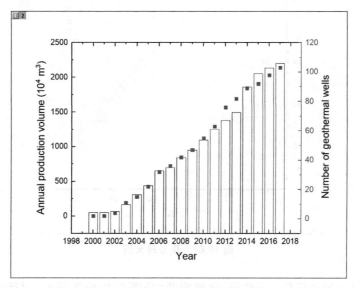

图 10.62 修改文字

7) 局部放大图

步骤①：单击菜单栏"文件"→"导入"导入数据到 Origin 工作表(图 10.63)中，全选数据表，单击左下角的"折线图"图标，软件自动跳转到 Graph 窗口(图 10.64)。

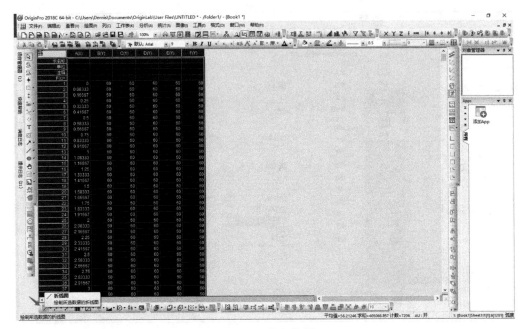

图 10.63　导入数据

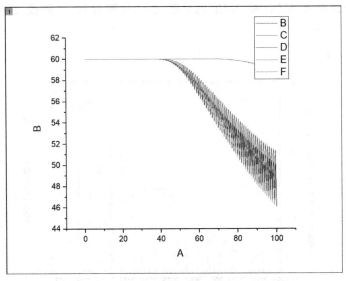

图 10.64　局部放大图

步骤②：双击图中任意一条曲线，在弹出的"绘图细节-绘图属性"对话框中，在"组"选项卡（图 10.65）中将"编辑模式"设置为"独立"，单击并进入"线条"选项卡（图 10.66），将"……A(X),B(Y)……"的线条"宽度"设置为 3，"颜色"设置为"浅灰"；将"……A(X),C(Y)……"的线条"宽度"设置为 1，"颜色"设置为"灰"；将"……A(X),D(Y)……"的线条"宽度"设置为 0.5，"颜色"设置为"深灰色"；将"……A(X),E(Y)……"的线条"宽度"设置为 0.5，"颜色"设置为"红"；将"……A(X),F(Y)……"的线条"宽度"设置为 0.5，"颜色"设置为

"蓝",单击"确定"按钮。

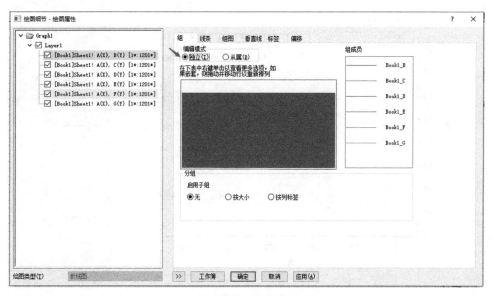

图 10.65 "组"选项卡

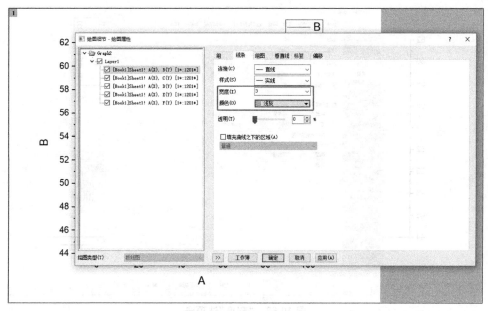

图 10.66 "线条"选项卡

步骤③:双击 Graph 窗口中的任意空白位置,在弹出的"绘图细节-页面属性"对话框中,在"打印/尺寸"选项卡(图 10.67)中,将尺寸"宽度"设置为 400,"高度"设置为 200,单击"确定"按钮。

步骤④:双击 X 轴,在弹出的"X 坐标轴-图层 1"对话框中,将"刻度"选项卡(图 10.68)中的"起始"设置为 0,"结束"设置为 100,将"主刻度"的"值"设置为 10。

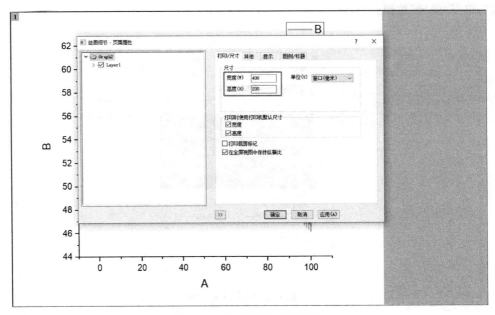

图 10.67 "打印/尺寸"选项卡

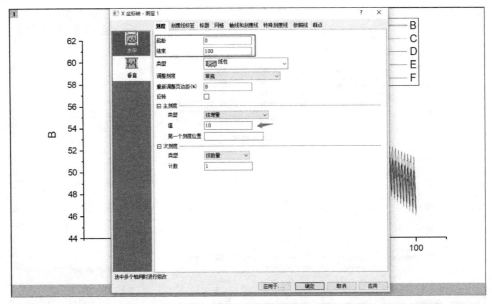

图 10.68 "刻度"选项卡

步骤⑤：单击并进入"轴线和刻度线"选项卡（图 10.69），将"下轴"和"左轴"的"主刻度"与"次刻度"的"样式"设置为"朝内"。

步骤⑥：单击"上轴"图标，选中"显示轴线和刻度线"以及"对下轴和上轴使用相同的选项"（图 10.70），并按照同样的方法设置"右轴"，单击"确定"按钮。

步骤⑦：单击左边工具栏中的"直线工具"图标，按住 Shift 键，在图中绘制一条直线，双击该直线，在弹出的"对象属性-Line"对话框中，将"线条"选项卡中的"宽度"设置为 1,"类

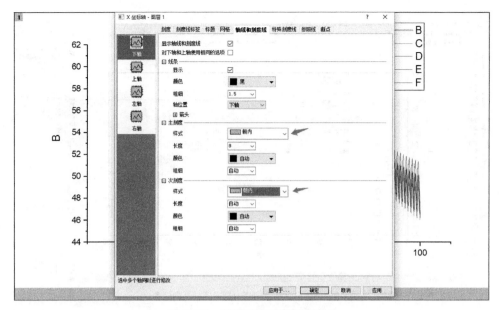

图 10.69 "轴线和刻度线"选项卡

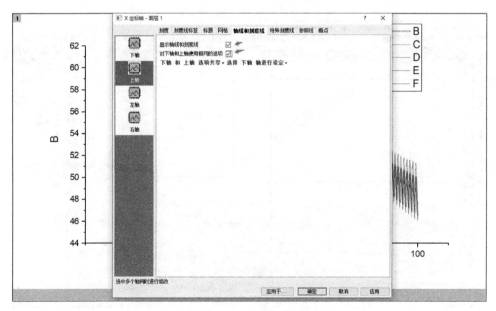

图 10.70 设置轴线

型"设置为"划线",单击"确定"按钮(图 10.71)。

步骤⑧:单击左边工具栏中的"放大"图标,按住键盘上的 Ctrl 键,并按住鼠标左键拖动需要局部放大的范围,当确定范围后松开鼠标左键,软件会自动生成 Enlarged 窗口,删除"图例"及"X 轴"和"Y 轴"的标题文字,并按照前面的方法对局部放大图进行相关设置(图 10.72,图 10.73),通过快捷键 Ctrl+J 复制 Enlarged 窗口中的图形,返回 Graph 窗口,通过快捷键 Ctrl+V 粘贴复制好的图形,通过拖拽和缩放操作将局部方法图放到合适的位置。

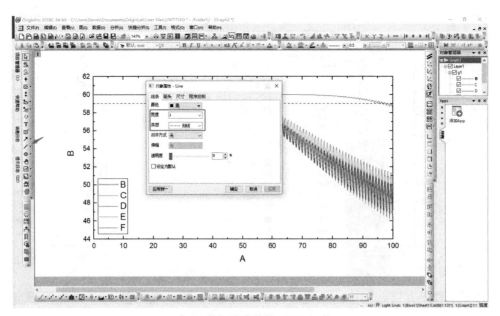

图 10.71 绘制直线

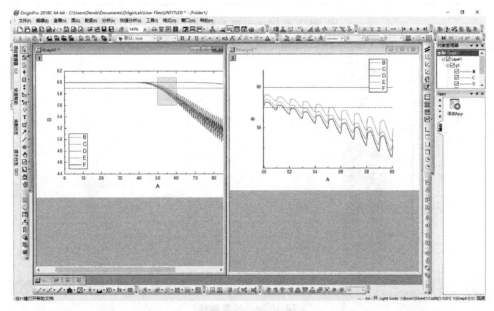

图 10.72 设置图形 1

步骤⑨：分别双击"Y 轴"和"X 轴"的标题文字，按图 10.74 修改坐标轴的标题，双击"图例"文字，在弹出的"文本对象-Legend"对话框中，将"字体大小"设置为 18，并按图 10.74 修改图例的文字内容，单击"确定"按钮，即可完成绘图（图 10.75）。

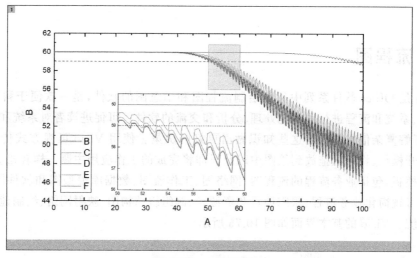

图 10.73　设置图形 2

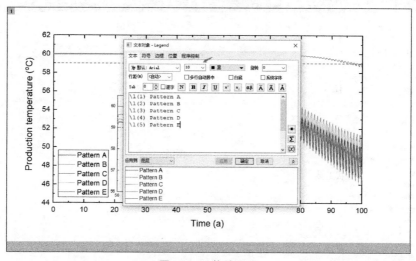

图 10.74　修改文字

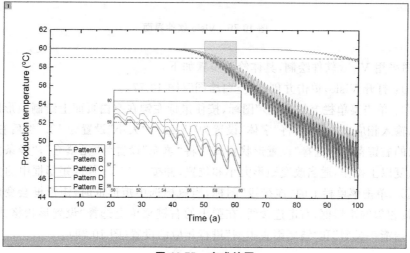

图 10.75　完成绘图

10.2 流程图

Visio 是 Office 软件系列中负责绘制流程图和示意图的软件,是一款便于科研人员就复杂信息、系统和流程进行可视化处理、分析和交流的软件。可促进读者对系统和流程的了解,深入了解复杂信息并利用这些知识做出更好的决策。使用 Visio 可视方式传递重要信息就像打开模板、将形状拖放到绘图中以及对即将完成的工作应用主题一样轻松。Visio 提供了各种模板,包括业务流程的流程图、网络图、工作流图、数据库模型图和软件图,这些模板可用于可视简化业务流程、跟踪项目和资源、绘制组织结构图、映射网络、绘制建筑地图以及优化系统。Visio 的基本界面如图 10.76 所示。

图 10.76　Visio 软件界面

示意图采用 Visio 软件绘制,具体绘制步骤如下:

步骤①:打开 Visio,单击并创建"空白绘图"(图 10.77)。

步骤②:单击菜单栏上的"矩形"图标,按住鼠标左键在空白页面上绘制矩形图框,双击矩形图框,输入相应文本内容,将"字体"设置为"Arial","大小"设置为 12。然后右击矩形图框,在弹出的右键菜单中选择"设置形状格式",将"填充"设置为"无填充","线条宽度"设置为 1 磅,重复以上操作,适当改变矩形大小和位置,并按图 10.78 修改矩形框中的文字内容。

步骤③:单击菜单栏上的"连接线"图标,当鼠标选中矩形框时,矩形框会变成"绿色",拖拽并连接相邻的矩形框,右击连接线,在弹出的右键菜单上选择"设置形状格式",并对连接线的"开始箭头类型"和"结尾箭头类型"进行相应的设置(图 10.79)。

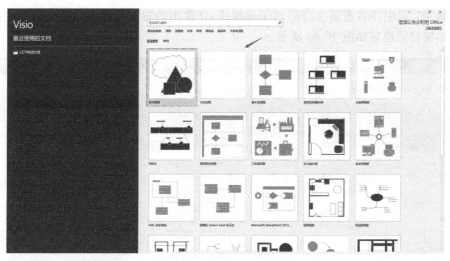

图 10.77　创建"空白绘图"

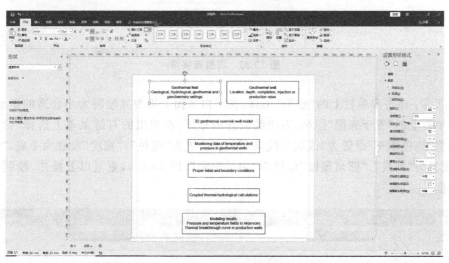

图 10.78　修改矩形框中文字

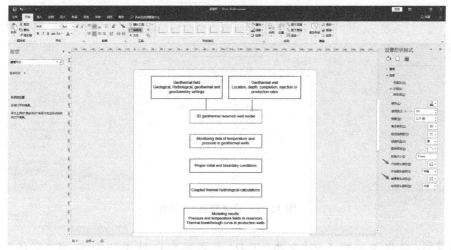

图 10.79　设置箭头类型

步骤④：选中并右击最上方的两条连接线，在弹出的右键菜单中，选择"直线连接线"，并调整连接线的位置如图 10.80 所示。

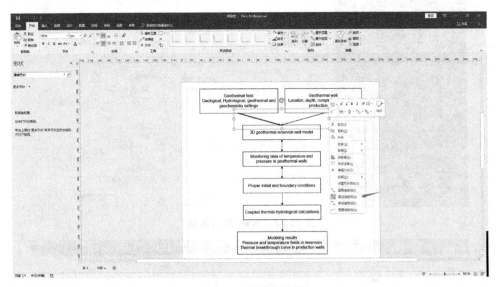

图 10.80　调整连接线

步骤⑤：单击菜单栏上的"矩形"工具，在空白页面上拖拽并绘制大小合适的矩形；然后单击菜单栏上的"置于底层"图标，右击绘制好的矩形，在弹出的右键菜单上选择"设置形状格式"，将"纯色填充"设置为"浅灰"，线条"颜色"设置为"蓝色"，"宽度"设置为 1 磅，"短划线类型"设置为"划线"，"圆角预设"设置为"中等圆角"（图 10.81），重复以上操作，绘制其余两个矩形框。

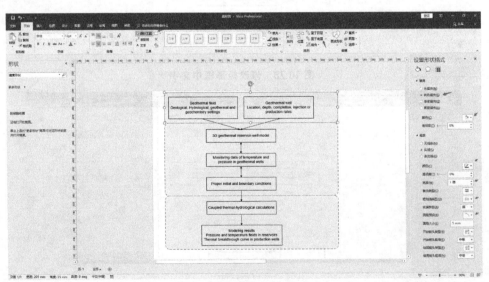

图 10.81　设置颜色、宽度等

步骤⑥：单击菜单栏上的"文本"工具，在页面空白处添加文本，将"字体"设置为

"Arial","字体大小"设置为 18,并将文本框拖拽到合适位置,即可完成绘图(图 10.82)。

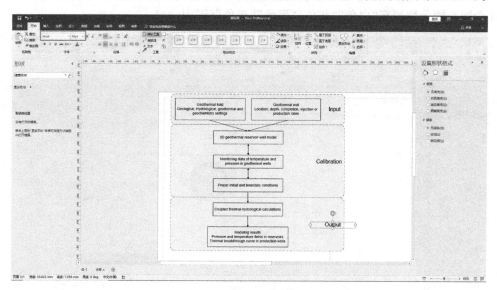

图 10.82　设置字体

10.3　示意图

示意图采用 Visio 软件绘制,具体绘制步骤如下:

步骤①:打开 Visio,单击并创建"空白绘图"(图 10.83)。

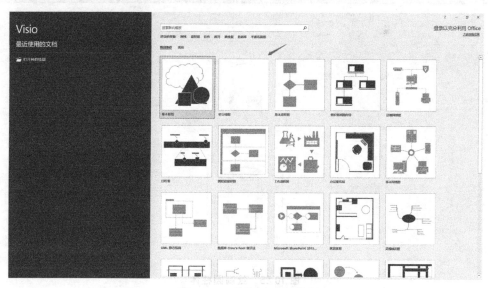

图 10.83　创建"空白绘图"

步骤②:单击并拖拽水平和垂直标尺参考线至空白页面的合适位置,单击菜单栏中"椭圆"工具,按住 Shift 键,拖拽并绘制 4 个半径不等的同心圆,并单击菜单栏中的"填充"工具,给 4 个同心圆分别填充"白色""灰色""黑色""蓝色"(图 10.84)。

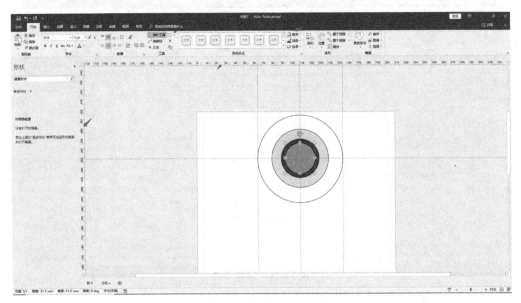

图 10.84　绘制同心圆并填色

步骤③:拖拽并添加更多的标尺参考线,以便绘制圆柱体,灵活运用"椭圆"和"直线"工具,可以快速绘制出如图 10.85 所示的圆柱体,将不可见的部分"短划线类型"设置为"划线",圆柱体底部颜色填充的"透明度"设置为 50%。

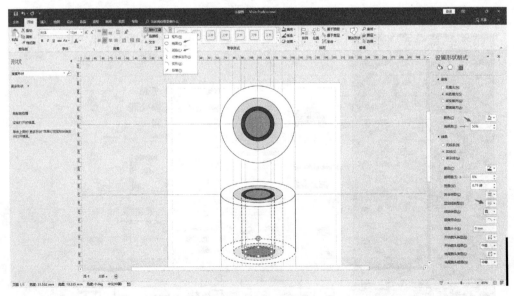

图 10.85　绘制圆柱体

步骤④:单击菜单栏中的"弧形"工具,为圆柱体底部添加 2 条弧线,并将线条"宽度"设置为 1 磅(图 10.86)。

第 10 章　图片制作指南

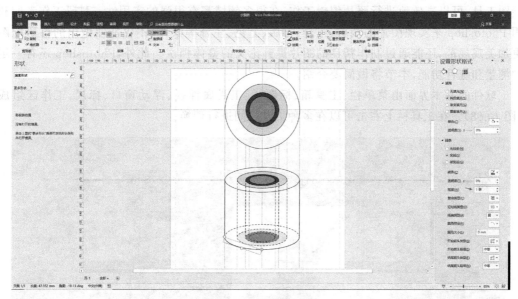

图 10.86　设置宽度

步骤⑤：灵活运用"线条""弧形""文本"等工具，给示意图添加不同的标注，并根据需要设置"线条"两端的箭头，即可完成绘图（图 10.87）。

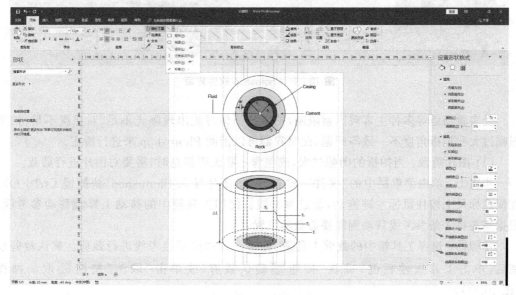

图 10.87　完成绘图

10.4　图像

Photoshop 简称"PS"，是由 Adobe 公司开发和发行的图像处理软件，Photoshop 主要处理由像素点所构成的数字位图，兼有简单的矢量和动画处理能力。得益于其丰富的编修与绘图工具,可以高效地进行照片编辑工作,在很多领域都有很强的实用性。科研工作者们利用 Photoshop 来实现图片的裁剪、组合、排版、上色、注释等操作,不仅能使做出的科研配图更加美观规范,还能通过制作精良的图片提升论文的整体档次。此外,Photoshop 还有一定的测量和统计功能,本节将做简要介绍。

软件的基本界面由菜单栏、工具箱、标题栏、工具属性栏、浮动窗口、标尺、工作区组成(图 10.88)。在工具栏上右击可以在多种工具之间进行切换。

图 10.88　Photoshop 软件界面

受实验场地等多种主客观因素限制,拍摄的图像可能出现曝光不足、对比度不足、拍摄图幅过大和拍摄角度不一致等问题,此时我们可以借助 Photoshop 来进行矫正。

(1) 图片剪裁:当拍摄的图幅过大,或包含一些无用信息时,需要对图片进行剪裁。

步骤①:单击菜单栏中的"文件"→"打开"将图片导入 Photoshop(快捷键 Ctrl+O)。在横纵标尺上按住鼠标左键拖动,新建参考线。使用工具箱中的移动工具 ![] 移动参考线,根据需要将 4 条参考线移动到需要剪裁的位置。

步骤②:选择工具箱中的裁剪工具 ![],拖动图片边缘至参考线进行裁剪。确认裁剪范围后,单击工具属性栏的"确认"按钮 ![] 确认裁剪,或单击"取消"按钮 ![] 取消操作(图 10.89)。

另外,相机位置不在被拍摄物体的正上方时,会由于透镜成像"近大远小"的原理导致被拍摄物体产生几何畸变(如长方形物体在照片中为平行四边形或梯形)。此时可以使用透视裁剪工具 ![] 进行矫正(图 10.90)。裁剪时拖动参考网格与被拍摄物体边缘平行,裁剪完成后物体内部的透视关系也会自动调整。

(2) 图片增强:受限于研究现场的光照和天气条件,图片可能出现不够明亮、色彩不鲜艳、模糊等问题,都可以通过 Photoshop 自带的工具进行增强。

第 10 章 图片制作指南

图 10.89　裁剪图片

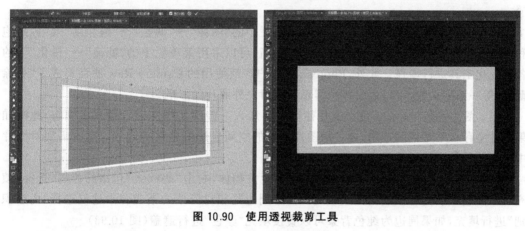

图 10.90　使用透视裁剪工具

如果拍摄时光照不足，快门与光圈配合不合理，可能使图像变得"暗淡""发黑"，这时可以使用菜单栏中的"图像"→"调整"→"曝光度"或"亮度"进行增强（图 10.91）。需要注意的

图 10.91　调整曝光度

是,曝光度和亮度的概念有所区别,在 HSB 模式下调整亮度时,饱和度 S 不变,仅亮度 B 变化;调整曝光度时,饱和度 S 和亮度 B 同时变化。

拍摄时光源比较单一,可能会导致图片内出现阴影,此时阴影位置曝光不足,而接受光照的位置比较明亮清晰,此时不能整体调整曝光。需要使用菜单栏中的"图像"→"调整"→"阴影/高光"进行调整(图 10.92)。

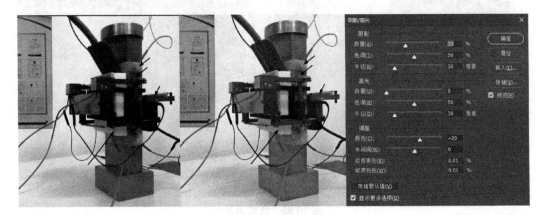

图 10.92 调整光影

如果图片色彩不够"艳丽"可以使用菜单栏中的"图像"→"调整"→"饱和度"或"自然饱和度"进行调整;如果图片模糊或边缘不够清晰,可以采用菜单栏中的"滤镜"→"锐化"下的各种锐化工具进行增强。此外,Photoshop 用户广泛使用的 Camera Raw 滤镜集成了"去除薄雾"等一系列功能,可以用于修正雾霾天气对户外照片的影响。

(3) 图片填充和标注:论文图片往往需要加入一些图文标注以方便读者理解,此时如果照片中存在不必要的杂物,会占用图片标注的空间且影响整体美观性。此时需要对图片中的无关物体进行遮盖填充。

步骤①:使用多边形套索工具选择出无关物体,右击"蚂蚁线"包络的范围,选择"填充"→"颜色"或"内容识别",完成填充。如果周边是相对简单的渐变背景建议选择"内容识别"进行填充,如果周边为纯色背景可以直接填充"颜色"进行遮盖(图 10.93)。

图 10.93 颜色"填充"

步骤②:图片中需要叠加其他图片进行说明时,单击菜单栏"文件"→"置入嵌入的智能对象"选择需要导入的图片,完成导入。使用工具箱中的移动工具 可以改变导入图像的位置和大小(图 10.94),拖动边缘改变大小时按住 Shift 键可以保持其长宽比不变。

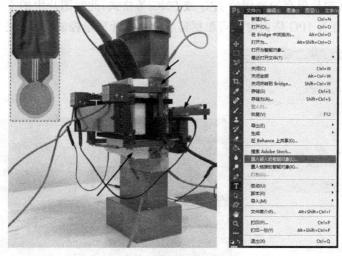

图 10.94 调整图像位置、大小

步骤③:使用工具箱中的自定义形状工具 绘制箭头;使用工具箱中的文字工具 添加文字说明(图 10.95)。箭头和文字的相关属性可以在相应的工具属性栏中进行调整。

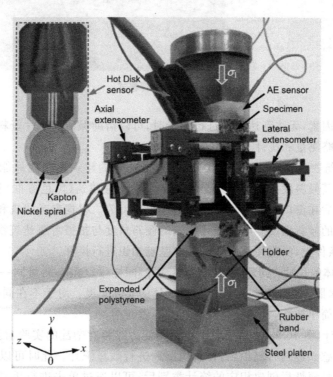

Fig. 1. A typical specimen setup showing positions of a pair of extensometers, AE sensors and the Hot Disk sensor.

图 10.95 添加文字说明

（4）导出和保存：制作完成后单击"文件"→"保存"将图片保存为后缀为.psd 的工程文件，此后用 Photoshop 打开可以对其中独立的文字和图层进行二次编辑。还可以单击"文件"→"另存为"将图片保存为通用的 JPG 或 PNG 等格式，此时图片中的文字等不可进行二次编辑。文件保存为 JPG 格式时还可根据需要调整图片质量控制文件大小（图 10.96）。

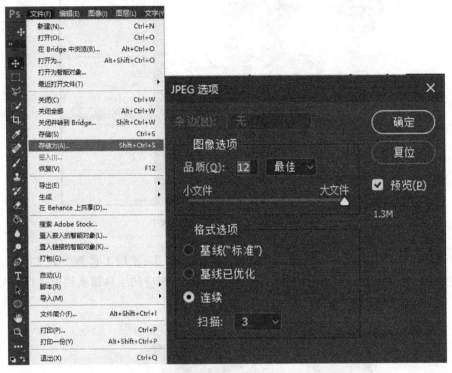

图 10.96　保存格式

受限于客观因素，无法到达现场进行测量时，可以借助保留的图片资料对角度、长度、面积等信息进行粗略测量和统计。

（1）测量建筑物倾斜度（图 10.97）：对于已经倒塌的建筑物往往只能通过倒塌前的影像资料粗略估算其倾斜度。使用标尺工具 分别沿参考建筑（黑色方框）和倾斜建筑（白色方框）画直线，然后在窗口顶部工具属性栏上读出角度值。根据参考建筑和倾斜建筑的角度差估算倾斜建筑的倾斜角度。测量照片得到的倾斜角度与拍摄者跟建筑物倾斜方向的相对位置相关，如果条件允许，应使用三角函数等基础知识进行换算。

（2）测量物体的长度：同理，使用标尺工具 同样可以粗略测量同一平面内两物体的相对大小，根据参照物的实际长度换算目标物体的实际长度。对于不在同一平面内的物体，不可以直接进行简单换算，具体计算方法此处不做介绍。

（3）数量统计（图 10.98）：对于有一定运动能力或相对杂乱的实验对象，需要统计其数量时往往需要在感兴趣的时刻拍照，再利用照片进行数量统计。此时可以利用计数工具 进行统计。在工具属性栏创建相应的统计类别后，可以通过单击进行手动计数。工具属性栏最左侧显示统计的总数和当前类别计数。

图 10.97　测量建筑物倾斜度

图 10.98　数量统计

此外，对于其他科研场景，Photoshop 还可以对影像进行圆度、面积和灰度值的粗略测量，此处不做介绍。

10.5 地图

对于地理、地质类的专业图件,建议使用业内通用的专题制图软件(如 ENVI、ArcGIS、MapGIS、CorelDRAW 等)进行地图制作,在此不做介绍。对于其他学科,需要在公版地图或前人研究的基础上简单标注研究区范围和重点位置时,可以使用 Photoshop 进行图片处理。

步骤①:网络上获取的公版地图或前人论文中下载的地图文件多为位图,此时应检查图片的清晰度,如果清晰度达到发表要求,可以直接参照 10.4 节的内容按需对图片进行修改。

步骤②:如果图片清晰度过低,则使用菜单栏中的"图像"→"画像大小"调整画像大小和分辨率,一般图像宽度超过 160 mm、分辨率 600 dpi 可以达到清晰度要求(图 10.99)。

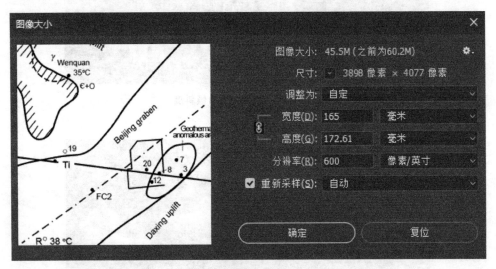

图 10.99 调整图像大小

步骤③:底图分辨率提高后用钢笔工具 对图片中需要的信息进行描摹,在工具属性栏中可以调整描摹时钢笔的颜色、描边形状、填充颜色(图 10.100)。

步骤④:如需添加标注,使用自定义形状工具 、矩形工具 、椭圆工具 、直线工具 和文字工具 进行标注。注意,使用矩形工具时按住 Shift 键和鼠标左键进行拖拽生成正方形;使用椭圆工具时按住 Shift 键和鼠标左键拖拽生成圆形;使用直线工具时按住 Shift 键和鼠标左键拖拽生成 0°或 90°的直线(图 10.101)。

步骤⑤:文件的保存和图片的导出,参考 10.4 节相关内容。

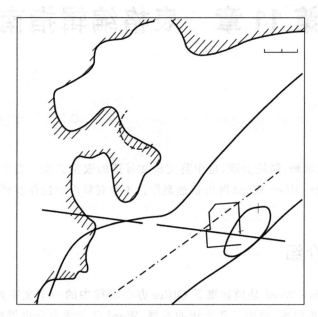

图 10.100 地图描摹

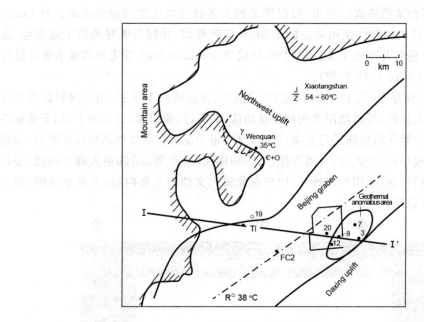

图 10.101 添加标注

第 11 章 表格编辑指南

"三线表"层次清晰、简洁美观,是中英文论文首选的表格类型。由于其结构较为简单,通常使用 Microsoft Office Word 即可快速制作。本章对软件和操作步骤进行简要介绍。

11.1 软件介绍

Microsoft Office Word 是微软旗下 Office 办公套件中的一个文字处理应用程序。自 20 世纪 80 年代问世以来,经历了几十年的发展,Word 已经成为全世界最为通用的办公和文字处理软件之一。它在兼容多种语言的排版习惯的同时,兼具简单的图片和表格处理能力,支持 PDF 等多种文档格式的导出,可以满足绝大多数科技论文的投稿要求。自 Office 2007 开始,办公套件内的软件使用全新的 Ribbon 交互界面,并拥有更好的向上兼容性,逐渐解决了低版本不能打开高版本文件的问题,建议使用 Office 2007 及更高的版本软件进行相关操作(本文以 Office 2016 为例)。

Office 2016 工作界面(图 11.1)由标题栏、功能区、快速访问工具栏、用户编辑区等部分构成:①快速访问工具栏,可以使用常用的快速功能,如保存、撤销等;②选项卡,用于显示各个功能的名称,对功能区的按钮进行分类;③标题栏,用于显示当前文档名称和扩展名;④窗口按钮,可以执行最小化、最大化、关闭等操作;⑤功能区,包含 Word 的绝大部分功能;⑥状态栏,用于显示文档信息;⑦用户编辑区,用户在此输入文档的主要内容;⑧显示比例,用于放大和缩小用户编辑区。

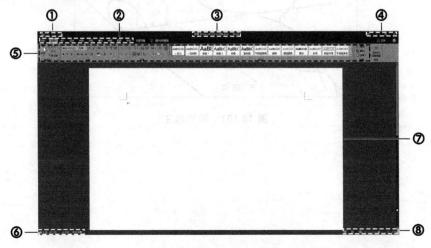

图 11.1　Office 2016 工作界面

11.2 操作步骤

步骤①：单击 Word 菜单中的"插入"→"表格"根据需要选择行数和列数（图 11.2）。

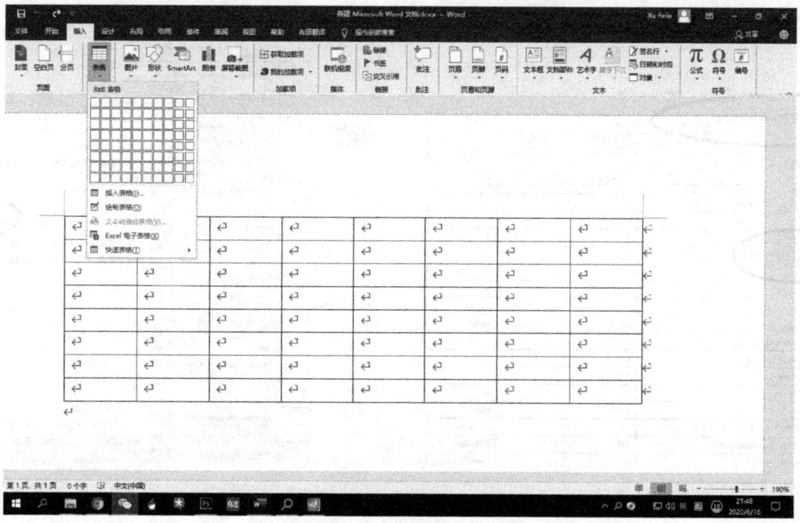

图 11.2　选择行数和列数

步骤②：将数据输入表格，行或列数量不够时可以右击表格选择"插入"→"插入行"或"插入列"进行添加（图 11.3）。

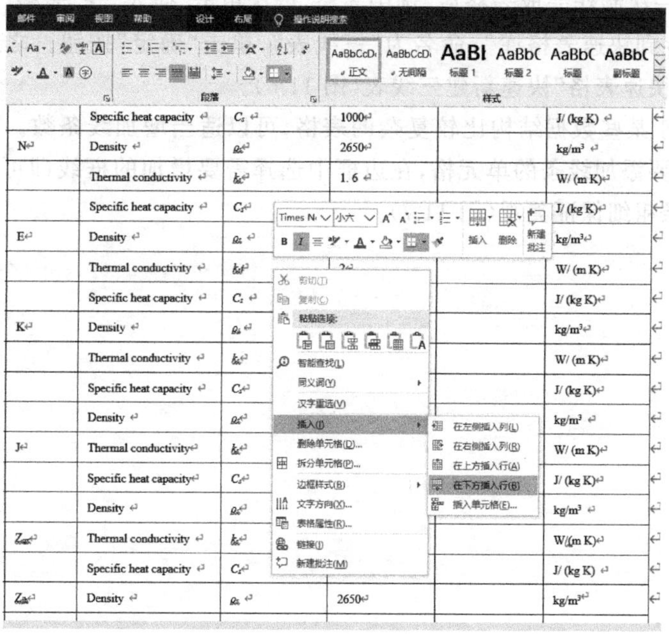

图 11.3　插入行或列

步骤③：选中需要合并的单元格，右击选择"合并单元格"。在"段落"和"字体"选项卡中分别选择合适的字体大小和行间距，尽量保证表格可以在同一页面内显示（图 11.4）。

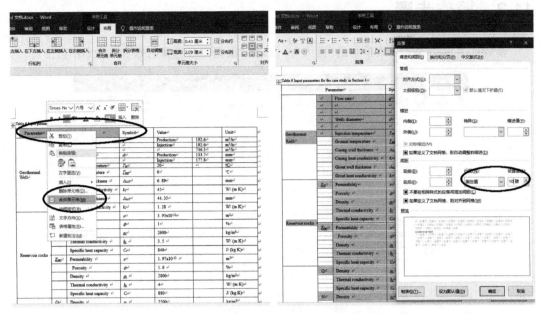

图 11.4　合并单元格

步骤④：选中表格，单击设计，在边框中的笔画粗细选择 1.5 磅；在边框下拉框先选择无框线，再选择上框线和下框线；选中第一行后，在边框中选择下框线；得到三线表（图 11.5）。

步骤⑤：第一次制作完成表格后，选中表格，依次单击"插入"→"表格"→"快速表格"→"将所选内容保存到快速表格库"，命名为"三线表"，单击"确定"按钮。之后，就可以从"插入"→"表格"→"快速表格"快速新建三线表（图 11.6）。

步骤⑥：对于某些数据结构比较复杂的表格，可以适当增加线条数。操作方法与步骤④相似。选中需要添加线条的单元格，在边框中选择需要增加的框线即可。线条数量较多时，可以降低线条粗细保证美观（图 11.7）。

第 11 章 表格编辑指南

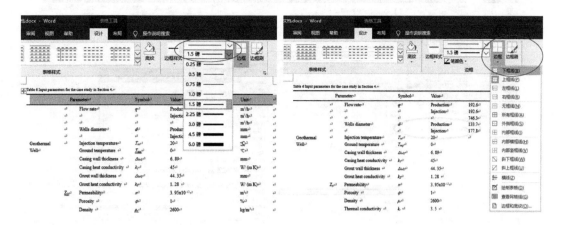

Table 6
Input parameters for the case study in Section 4.

	Parameter		Symbol	Value		Unit
Geothermal Well		Flow rate	q	Production	192.6	m³/h
				Injection	192.6	m³/h
					746.3	m³/h
		Wells diameter	d	Production	133.7	mm
				Injection	177.8	mm
		Injection temperature	T_{in}	20		°C
		Ground temperature	T_{top}	0		°C
		Casing wall thickness	Δw_1	6.89		mm
		Casing heat conductivity	k_1	45		W/(m K)
		Grout wall thickness	Δw_1	44.35		mm
		Grout heat conductivity	k_2	1.28		W/(m K)
Reservoir rocks	Zjt	Permeability	κ	3.95×10^{-12}		m²
		Porosity	ϕ	1		%
		Density	ρ_r	2600		kg/m³
		Thermal conductivity	k_r	3.5		W/(m K)
		Specific heat capacity	C_r	840		J/(kg K)
	Zjw	Permeability	κ	1.97×10^{-12}		m²
		Porosity	ϕ	1.6		%
		Density	ρ_r	2800		kg/m³
		Thermal conductivity	k_r	4		W/(m K)
		Specific heat capacity	C_r	880		J/(kg K)
Cap rocks	Q	Density	ρ_s	2500		kg/m³
		Thermal conductivity	k_s	1.5		W/(m K)
		Specific heat capacity	C_s	1000		J/(kg K)
	N	Density	ρ_s	2650		kg/m³
		Thermal conductivity	k_s	1.6		W/(m K)
		Specific heat capacity	C_s	850		J/(kg K)
	E	Density	ρ_s	2700		kg/m³
		Thermal conductivity	k_s	2		W/(m K)
		Specific heat capacity	C_s	900		J/(kg K)
	K	Density	ρ_s	2600		kg/m³
		Thermal conductivity	k_s	2.5		W/(m K)
		Specific heat capacity	C_s	850		J/(kg K)
	J	Density	ρ_s	2700		kg/m³
		Thermal conductivity	k_s	2.3		W/(m K)
		Specific heat capacity	C_s	950		J/(kg K)
	Zqx	Density	ρ_s	2600		kg/m³
		Thermal conductivity	k_s	1.7		W/(m K)
		Specific heat capacity	C_s	920		J/(kg K)
	Zjh	Density	ρ_s	2650		kg/m³
		Thermal conductivity	k_s	1.8		W/(m K)
		Specific heat capacity	C_s	850		J/(kg K)
Faults		Permeability	κ_{fr}	4.93×10^{-7}		m²
		Thickness	d_{fr}	10		cm

图 11.5 制作三线表

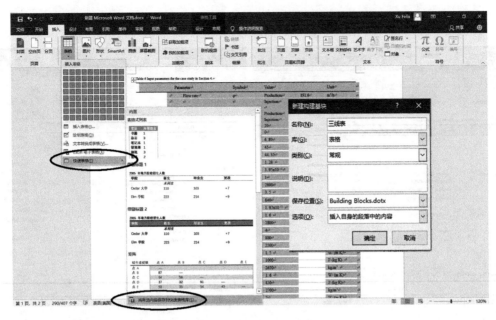

图 11.6 快速新建三线表

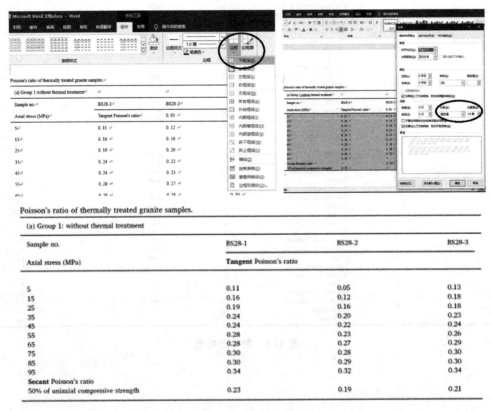

图 11.7 增加表格线条

第 12 章 常用句式与词汇

12.1 引言

1) 研究背景

以下句式可用于描述研究工作的重要性。

Fire-induced spallation of rock has also been ***identified as a growing*** concern in tunnel and mine safety.[40]

Fluid injection in deep sedimentary unites with hydraulic connectivity to basement faults ***has been attributed to inducing*** earthquakes in numerous case studies in ×××.[41]

Conclusively distinguishing human-induced earthquakes ***solely on the basis of*** seismological data ***remains challenging***.[42]

The first Dutch geothermal sites proved that Hot Sedimentary Aquifers exploitation can ***play an important role*** in a future low-carbon energy mix.[43]

It is therefore ***vital/essential*** to study the effect of temperature on tensile strength for designing structures within rock.[44]

The mechanism that drives the human-induced geohazards still ***eludes explanation***, and ***uncovering*** this mechanism ***relies heavily on*** our understanding of the failure characteristics of fractured rock.[45]

2) 研究现状

以下句式可用于客观评价前人工作。

At present, models of spall production ***remain largely empirical*** in nature or adopt simplifying assumptions that ignore microstructural heterogeneity.[46]

The impact of fracture normal strength and topography on the hydro-mechanical behavior in a fracture has therefore been ***thoroughly scrutinized***…[47]

Extensive research has been conducted to understand the effect of heat treatment on the tensile strength of rocks.[48]

以下句式用于指出研究不足,顺势引出所要研究的问题。

While these recent works illustrate ***the first attempts to bridge the gap between*** nanoscale friction measurements and fault dynamics, ***many questions still remain open***.[49]

However, the impact of network structure and hydraulic variability on cement grout propagation in water-saturated fracture networks ***remains an open issue***.[50]

Numerous stimulation tests have been performed on ××× projects during the past three decades, however, ***there is much room for improvement*** in our knowledge and understanding of the mechanisms of ×××.[51]

Even though literature on the mechanical behavior of the rock fractures with infill material ***is quite extensive***, ***a rational framework*** has not been developed for assessment of the effects of stress state and degree of saturation of infill material on their shear behavior.[52]

Very limited research has been carried out on the tensile strength of sedimentary rocks when exposed to high temperatures.[48]

While the performance of solute tracers in providing aquifer parameter, or flow rate estimates for heterogeneous aquifers has been ***extensively investigated*** using both field and numerical methods, investigations into the influence of aquifer heterogeneity on thermal transport and the interpretation of heat as a tracer are ***less common***.[53]

While ***a vanishingly small number of*** papers have simultaneously considered heat and solute tracers to explore groundwater systems, they have not done so ***in a sufficiently generalized way*** to ***draw reasonably generalized conclusions*** about how these tracers compare and contrast when specifically utilized for the objective of groundwater flow rate estimation in heterogeneous aquifers.[53]

3）研究目标

以下句式用于描述研究目的。

In this paper, we ***outline a numerical modeling effort*** investigating the grain-scale mechanics of thermal spallation.[54]

In this study, we present ***state of the art*** experimental results regarding the influence of the state of stress and of the injection rate on the onset of fault reactivation.[55]

12.2 方法

1）试验方法

以下句式用于描述试验材料。

The Eagle Ford shale is a Late Cretaceous-age sedimentary rock ***sourced*** from South Texas with highly variable composition, low porosity (5%～7%), high strength (uniaxial compressive strength ＞120 MPa) and very low permeability parallel to bedding planes.[56]

以下句式用于描述试验仪器与方法。

Each sample was ***jacketed*** using a 0.75 mm heat-shrink Viton sleeve, which was then ***sealed*** on two steel core holders using steel wires.[56]

The experiment was ***conducted*** using a direct shear apparatus (RLJW-2000).

The apparatus ***consists of*** two orthogonally arranged pistons which are ***supported*** by steel posts and beams with a sufficiently high stiffness (> 5 GN/m), and the pistons are ***servo-controlled*** to separately serve the prescribed loading and displacement rates.[57]

Normal and shear loads were ***measured*** through load cells (DOLI, Germany) with capacities of 180 and 120 kN, respectively. The accuracy of the applied load was ± 1% *of the full scale*.[57]

2）模拟方法

以下句式用于描述研究方法的选择依据与假设条件。

Indeed for many applications it may be ***prohibitively expensive*** to conduct appropriate experimental investigation under the appropriate conditions or impossible to obtain the relevant experimental data. ***In such an environment***, explicit small-scale numerical simulations ***can be used to provide insight into*** the fundamental processes contributing to thermal spallation.[40]

Here we assume that within the fracture, advection usually ***prevails on*** water thermal diffusivity, so that, thermal lag time is poorly influenced by conduction in water compared to conduction in rock.[58]

After thorough investigation, most of the more advanced machine learning algorithms were found to be ***more complex than required*** for this problem and therefore less suitable solutions, since ***simplicity is a welcomed attribute for a solution***.[59]

以下句式用于描述模拟方法与模拟工况。

We ***first compare*** the simulation results from the 2.5D continuum model ***with*** the experimental observation. To demonstrate the impacts of the altered layer on solute transport and subsequent geochemical reaction, we included ***comparative simulations with and without*** the diffusion limitation imposed by the altered layer on the fast-reacting mineral, calcite in this case.[60]

12.3 结果

以下句式用于描述图表内容。

Figure ××× ***shows/illustrates*** a comparison of the evolution in fracture permeability for the various fracture properties and stimulation scenarios.[61]

Figure ××× ***represents*** the initial fracture permeability distribution …[61]

The fracture permeability distribution is ***presented*** in Figure ×××.[61]

As fracture spacing increases, the chilled region adjacent to the injector is reduced in size (***see*** Figure ×××).[62]

Figure ××× shows the ***evolution*** of the relative stress intensity factor ***versus*** displacement, ***as a function of*** temperature treatment.[63]

以下句式用于描述研究结果的数据规律。

The GRS fracture permeability **declines** during shearing while an increased sliding velocity **reduces** the rate of permeability decline.[64]

This would **significantly increase** the computational times for the numerical simulation but only **marginally increase** those of the prediction model.[65]

The load-strain plot **remained linear** until the yield point and then the slope of load versus horizontal strain **decreased sharply**.[66]

12.4 讨论

以下句式用于研究结果之间的对比。

It is now **widely recognized** that the failure process is initiated by the onset ... **This is also true** in a traditional Brazilian test and the flattened Brazilian tests ...[66]

...crack initiation occurs at a stress level of around 40%～50% of the peak load which **is consistent with** the findings by Nicksiar and Martin (2013) for low porosity crystalline rock.[66]

The logarithmic value of the permeability decreases as effective stress increases, which is **in good agreement with** an exponential function (equation (3)).[56]

These observations are **in close agreement with** the previous results obtained on Westerly granite.[63]

以下句式用于指出需在未来继续深入研究的问题。

The fit **could certainly be improved** if the tensile strength data was also used for establishing the Hoek-Brown envelope.[66]

However, **further work is required to fully understand** the mechanisms responsible for the decrease in shear strength observed with increasing scale.[67]

Some further clarification must be made between roughness and shear strength scale effects.[67]

These questions cannot be addressed without more experimental results from the field and additional numerical modeling to **further explore** the effect of fracture geometry on fluid flow and fracture deformation, and the link to the seismic response of a fracture.[68]

12.5 词汇

以下列出若干可在学术论文使用的句式和动词。

It is an efficient way to **unmask** the mechanisms[69]

This strategy **eliminates** some major regions, where effects of field-wide injection are difficult to **unravel**.[70]

Since literature searches are seldom ***infallible***, the author apologizes to those researchers whose works have not been reported here.[71]

We also ***explore the broader application of the model*** at the end of this section.[72]

The model ***captures*** the progressive decrease in the dissolution rate of the fast-reacting minerals in the altered layer.[73]

第 13 章　投稿流程简介

论文投稿是学术研究工作中关键的一环,它关系着每一位科研工作者的命脉。通常,再好的研究工作如果没有遵循正确的投稿方法,那它依旧不能被国内外同行专家所接收。因此,掌握并熟知正确的投稿方法可保证自己的研究成果能早日发表。

13.1　前期准备(Preparation)

1) 选定期刊(Journal)

按照论文稿件内容及创新点选择合适期刊,并在期刊官网上仔细阅读作者指导(guide for authors)中的每一条信息,认真按照该期刊所要求的标题、作者信息、摘要、正文、图片、表格、致谢、参考文献等格式对拟投稿件进行全面格式检查。本章以爱思唯尔旗下的国际岩石力学与采矿科学期刊 *International Journal of Rock Mechanics and Mining Sciences* (Q1,IF=3.89)进行示例说明(图 13.1)。

图 13.1　期刊信息查询页面

2) 投稿信(Cover letter)

投稿信不是必须提供,且也无格式要求。但建议您提供投稿信,须写明文章类型、工作重要性及 3~4 个亮点(highlights),推荐审稿人及信息,回避审稿人及信息,以及文章通信作者及信息等。具体内容模板可参考 4.2 节。

3) 论文稿件(Manuscript)

根据期刊模板(Template)对稿件格式进行修改,包括题目、所有作者信息(姓名、单位、工作地址、邮编等)及作者顺序(第一作者、并列作者、通信作者等,各作者右上角应标明数字以对应其个人信息,而通信作者上角标还需标记*)。稿件中还应包含摘要(Abstract)、关键字(Keywords)、正文、致谢(Acknowledgements)、基金(Funding)和参考文献(References)。参考文献格式应严格按照期刊要求进行修改(IJRMMS 期刊要求文献格式符合 AMA 标准)(图 13.2)。此外,各期刊都有自己的参考文献模板,作者可从期刊官网下载模板并导入英文文献管理软件(包括 Mendeley、Endnote 等)中,其可自动对稿件中的文献格式进行批量修改。此外,稿件还应保证拼写和语法正确。其余要求应以 Guide For Authors P146 中规定为准。

```
Reference style
Text: Indicate references by (consecutive) superscript arabic numerals in the order in which they
appear in the text. The numerals are to be used outside periods and commas, inside colons and
semicolons. For further detail and examples you are referred to the AMA Manual of Style, A Guide for
Authors and Editors, Tenth Edition, ISBN 0-978-0-19-517633-9.
List: Number the references in the list in the order in which they appear in the text.
Examples:
Reference to a journal publication:
1. Van der Geer J, Hanraads JAJ, Lupton RA. The art of writing a scientific article. J Sci Commun.
2010;163:51–59. https://doi.org/10.1016/j.Sc.2010.00372.
Reference to a journal publication with an article number:
2. Van der Geer J, Hanraads JAJ, Lupton RA. The art of writing a scientific article. Heliyon.
2018;19:e00205. https://doi.org/10.1016/j.heliyon.2018.e00205.
Reference to a book:
```

图 13.2 参考文献格式

4) 图片(Figure)

图片应包含标题并按其在稿件中出现顺序进行标号,然后单独放在一个 Word 文件中,图片格式及分辨率应按 Guide For Authors 中要求进行处理。图片一般应为 TIFF(JPEG)或 EPS(PDF)格式,分辨率 DPI 应不小于 300(图 13.3)。其余要求应以 Guide For Authors 中规定为准。

```
Color artwork
Please make sure that artwork files are in an acceptable format (TIFF (or JPEG), EPS (or PDF), or
MS Office files) and with the correct resolution. If, together with your accepted article, you submit
usable color figures then Elsevier will ensure, at no additional charge, that these figures will appear
in color online (e.g., ScienceDirect and other sites) regardless of whether or not these illustrations
are reproduced in color in the printed version. For color reproduction in print, you will receive
information regarding the costs from Elsevier after receipt of your accepted article. Please
indicate your preference for color: in print or online only. Further information on the preparation of
electronic artwork.
```

图 13.3 图片格式要求

5) 表格(Table)

表格应包含标题并按其在稿件中出现顺序进行标号(图 13.4),然后单独放在一个 Word 文件中,其余要求应以 Guide For Authors 中规定为准。

```
Tables
Please submit tables as editable text and not as images. Tables can be placed either next to the
relevant text in the article, or on separate page(s) at the end. Number tables consecutively in
accordance with their appearance in the text and place any table notes below the table body. Be
sparing in the use of tables and ensure that the data presented in them do not duplicate results
described elsewhere in the article. Please avoid using vertical rules and shading in table cells.
```

图 13.4 表格格式要求

6) 利益冲突(Conflict of interest)

利益冲突是指在论文的出版过程中,所有参与者,包括作者、审稿人、编辑和编委会成员,如与文中所涉及的内容存在经济/个人关系,则需要在文章结尾处正式说明利益关系。作者在投稿过程中有责任说明可能引起文章发生偏倚的利益冲突(图 13.5)。具体模板可参考各期刊的 Guide For Authors P145。

Declaration of interest
All authors must disclose any financial and personal relationships with other people or organizations that could inappropriately influence (bias) their work. Examples of potential conflicts of interest include employment, consultancies, stock ownership, honoraria, paid expert testimony, patent applications/registrations, and grants or other funding. Authors should complete the declaration of interest statement using this template and upload to the submission system at the Attach/Upload Files step. If there are no interests to declare, please choose: 'Declarations of interest: none' in the template. This statement will be published within the article if accepted. More information.

图 13.5 利益冲突声明

13.2 稿件投递(Submission)

1) 注册投稿系统(Register)

若首次采用系统,则须单击 Register 进行注册。按照如下页面要求输入 E-mail address(邮箱)、First name(名)和 Family name(姓)等信息(图 13.6),信息填写完毕后点 Register 注册完毕,之后邮箱会收到账号注册成功的邮件。

Register
Please enter your details below
Already have an account? Please sign into EVISE

Email address *
Your title *
Dr
First name *
Family name *
Password *   Password guidelines
Confirm password *
Organization name *
Country / Region *

图 13.6 账号注册页面

2）登录投稿系统（Sign in）

输入 Username 和 Password 登录（图 13.7）。

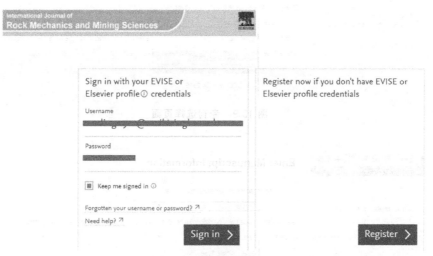

图 13.7　期刊登录页面

3）稿件投递（Submission）

输入账号密码，进入期刊投稿界面，单击 Start New Submission（投稿）（图 13.8）。

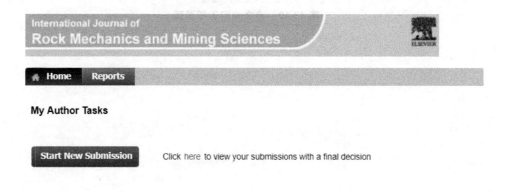

图 13.8　投稿开始页面

4）选择拟投文章类型（Regular issue/Special issue）

不同期刊系统会显示不同文章类型选项，如 Letter/Communication 等，作者应根据稿件类型进行选择，这里选择 Regular issue 下的 research paper（研究长文）进行后续投稿说明（图 13.9 和图 13.10）。

5）输入基本信息

依次输入标题（Title）、摘要（Abstract）、关键词（Topic）、作者（Authors）。须注意摘要文字限制。如文章有多个作者，须单击添加作者（Add author）后在新界面分别输入作者的职称（Title）、名字（Name）等信息（图 13.11）。勾选前方小框选择通信作者（Corresponding author）。

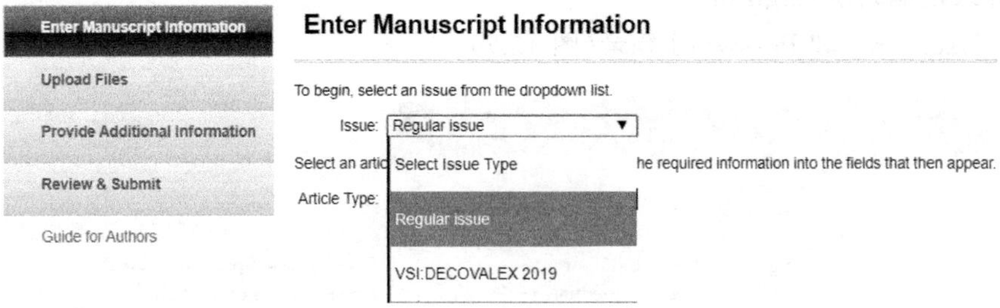

图 13.9　专刊选择页面

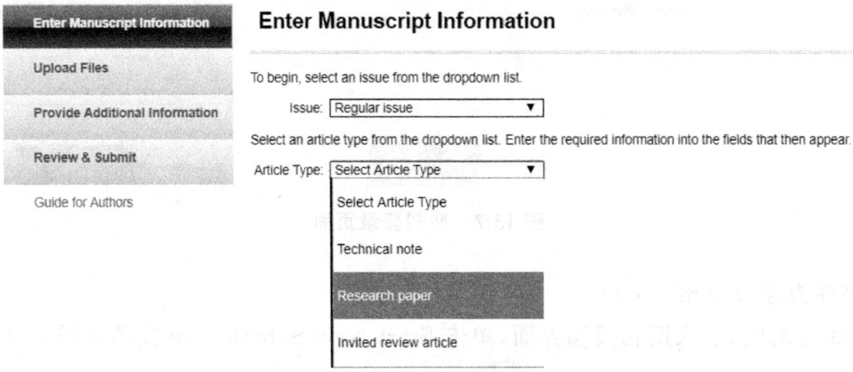

图 13.10　论文类型选择页面

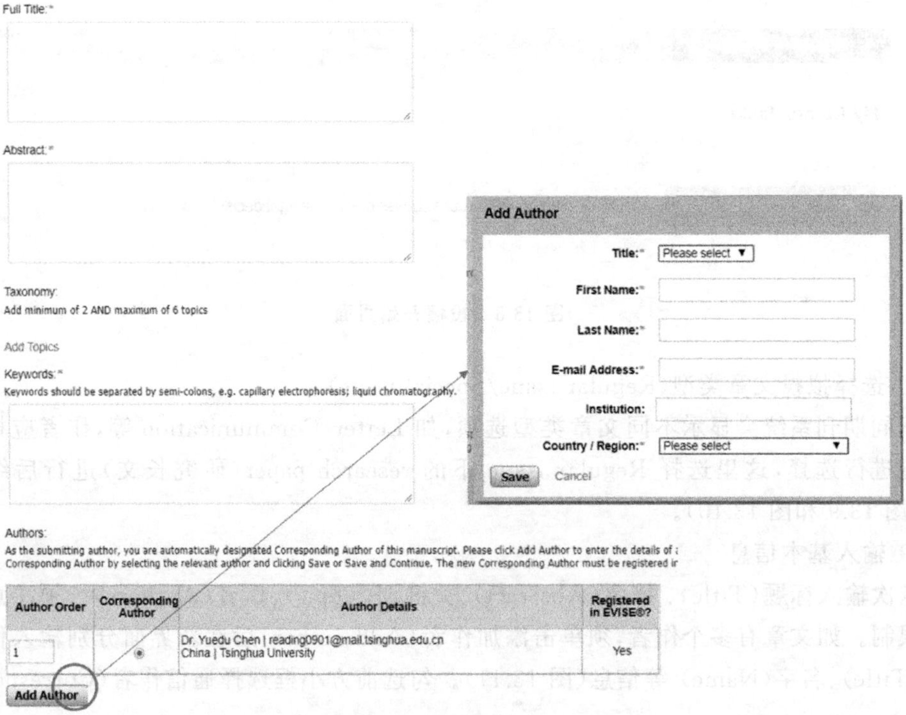

图 13.11　论文基本信息输入页面

6）上传附件

按照投稿指南要求，在下拉框中选择 Cover letter（投稿信）、Conflict of interest（利益冲突说明）、Manuscript（稿件）、Figure（图片）、Table（表格）等类型。其中 Cover letter 里须写明文章内容并进行亮点表述；而 Conflict of Interest 里须说明本文不存在一稿多投现象、稿件没有利益冲突。此外，还可上传其余辅助附件（图 13.12）。稿件全部上传完成后可用 Order 来调整顺序，通常排列顺序为 Cover Letter、Manuscript、Figure、Table、Conflict of interest。

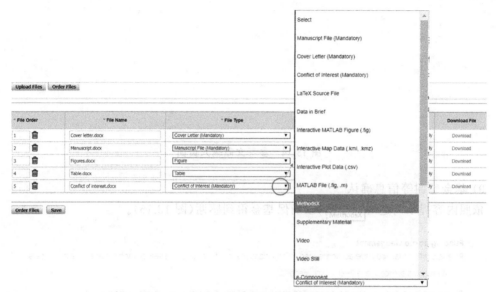

图 13.12　附件上传页面

7）分享研究数据（可选）

研究数据上传分享后有利于加强读者间的交流（图 13.13），可提高稿件的引用率和应用价值。

图 13.13　分享研究数据页面

8) 输入参考文献(References)

须将稿件正文中的 References 复制到这里(图 13.14)。须注意文献格式应与期刊投稿指南中所规定的文献格式一致。

图 13.14 参考文献输入页面

9) 基金感谢等信息确认

依照内容提示勾选对应小框,以确保基金得到感谢(图 13.15)。

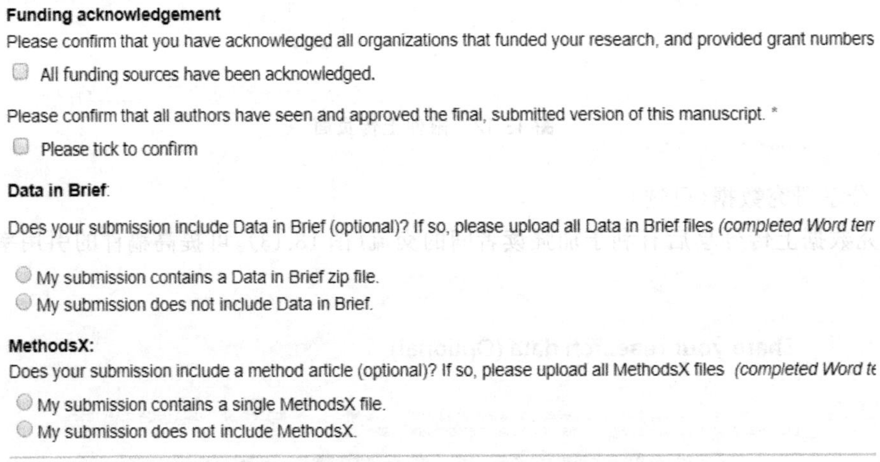

图 13.15 信息确认页面

10) 生成 PDF 确认文件

单击下载 Download PDF 后仔细查阅生成 PDF 格式有无错误,确认无误后单击 Complete Submission 完成稿件投递(图 13.16)。

11) 稿件投递成功邮件

投稿完成后,所有作者邮箱都会收到杂志社发来的稿件投递成功邮件,至此稿件投递工作全部完成(图 13.17)。

12) 查询稿件状态

回到投稿系统主页(Home)面后可查看稿件信息,包括题目(Title)、编号(No.)、状态

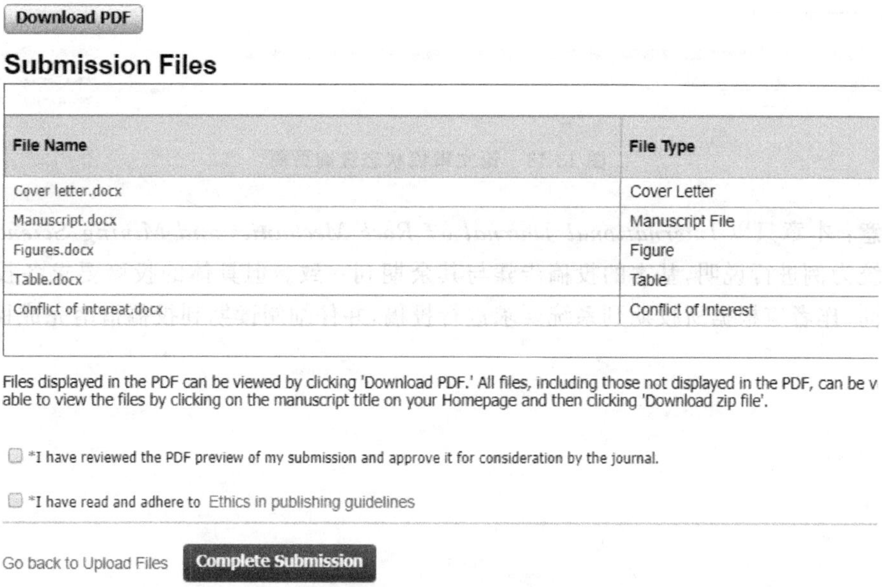

图 13.16　论文提交页面

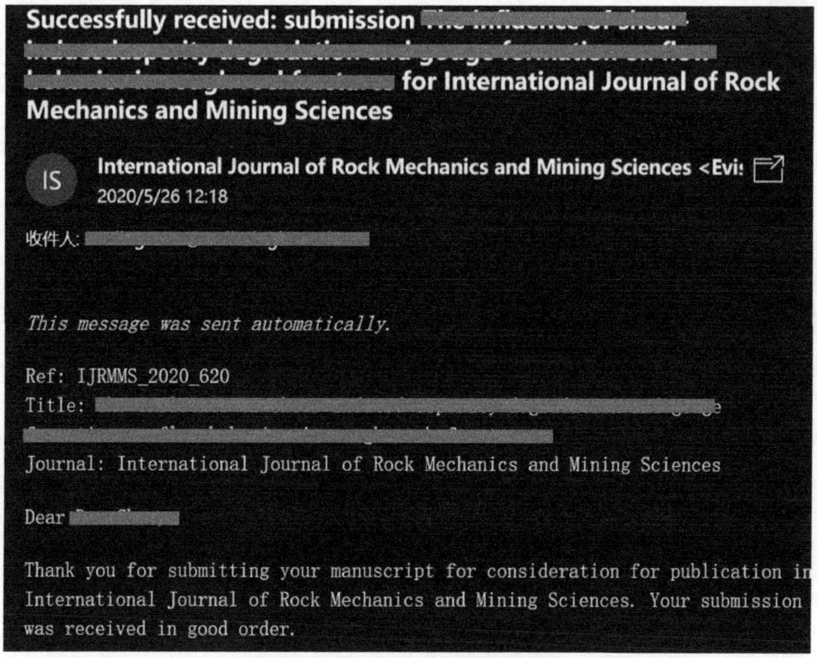

图 13.17　投稿成功通知邮件

(Current status)(图 13.18)。稿件状态会随审稿流程而发生变化,各期刊系统的状态名词会有不同。一般包括:Manuscript Submitted—With Editor—Reviewer Invited—Under Review—Decision in Process—Minor Revision/Major Revision/Reject and Resubmit/Reject—Accept 等。

图 13.18　论文审稿状态查询页面

注意：本章只以 *International Journal of Rock Mechanics and Mining Sciences* 期刊投稿系统为例进行说明，基本的投稿步骤与其余期刊一致。但具体的投稿要求各投稿系统会有不同，作者应根据所投期刊系统要求进行投稿，并仔细阅读期刊投稿指南完成稿件投递工作。

第 14 章　土木工程领域期刊缩写

土木工程领域期刊缩写见表 14.1。

表 14.1　土木工程领域期刊缩写

JOURNAL TITLE	JOURNAL ABBREVIATION	ISSN
土木工程（交叉学科）		
AUTOMATION IN CONSTRUCTION	AUTOMAT CONSTR	0926-5805
ADVANCES IN CIVIL ENGINEERING	ADV CIV ENG	1687-8086
BUILDING SIMULATION	BUILD SIMUL-CHINA	1996-3599
CANADIAN JOURNAL OF CIVIL ENGINEERING	CAN J CIVIL ENG	0315-1468
CIVIL ENGINEERING	CIVIL ENG	0885-7024
COASTAL ENGINEERING	COAST ENG	0378-3839
COMPUTER-AIDED CIVIL AND INFRASTRUCTURE ENGINEERING	COMPUT-AIDED CIV INF	1093-9687
ENGINEERING CONSTRUCTION AND ARCHITECTURAL MANAGEMENT	ENG CONSTR ARCHIT MA	0969-9988
ENGINEERING	ENGINEERING-PRC	2095-8099
EUROPEAN JOURNAL OF ENVIRONMENTAL AND CIVIL ENGINEERING	EUR J ENVIRON CIV EN	1964-8189
IRANIAN JOURNAL OF SCIENCE AND TECHNOLOGY-TRANSACTIONS OF CIVIL ENGINEERING	IJST-T CIV ENG	2228-6160
INTERNATIONAL JOURNAL FOR NUMERICAL METHODS IN ENGINEERING	INT J NUMER METH ENG	0029-5981
INTERNATIONAL JOURNAL OF ADAPTIVE CONTROL AND SIGNAL PROCESSING	INT J ADAPT CONTROL	0890-6327
JOURNAL OF ASIAN ARCHITECTURE AND BUILDING ENGINEERING	J ASIAN ARCHIT BUILD	1346-7581
JOURNAL OF CIVIL ENGINEERING AND MANAGEMENT	J CIV ENG MANAG	1392-3730
JOURNAL OF COMPUTING IN CIVIL ENGINEERING	J COMPUT CIVIL ENG	0887-3801
JOURNAL OF CONSTRUCTION ENGINEERING AND MANAGEMENT	J CONSTR ENG M	0733-9364
JOURNAL OF ENGINEERING DESIGN	J ENG DESIGN	0954-4828

续表

JOURNAL TITLE	JOURNAL ABBREVIATION	ISSN
JOURNAL OF MANAGEMENT IN ENGINEERING	J MANAGE ENG	0742-597X
JOURNAL OF OFFSHORE MECHANICS AND ARCTIC ENGINEERING-TRANSACTIONS OF THE ASME	J OFFSHORE MECH ARCT	0892-7219
JOURNAL OF PERFORMANCE OF CONSTRUCTED FACILITIES	J PERFORM CONSTR FAC	0887-3828
JOURNAL OF THE SOUTH AFRICAN INSTITUTION OF CIVIL ENGINEERING	J S AFR INST CIV ENG	1021-2019
JOURNAL OF WIND ENGINEERING AND INDUSTRIAL AERODYNAMICS	J WIND ENG IND AEROD	0167-6105
JOURNAL OF DYNAMICAL AND CONTROL SYSTEMS	J DYN CONTROL SYST	1079-2724
KSCE JOURNAL OF CIVIL ENGINEERING	KSCE J CIV ENG	1226-7988
MARINE GEORESOURCES & GEOTECHNOLOGY	MAR GEORESOUR GEOTEC	1064-119X
MATHEMATICAL PROBLEMS IN ENGINEERING	MATH PROBL ENG	1024-123X
NONDESTRUCTIVE TESTING AND EVALUATION	NONDESTRUCT TEST EVA	1058-9759
OCEAN ENGINEERING	OCEAN ENG	0029-8018
OCEAN MODELLING	OCEAN MODEL	1463-5003
PERIODICA POLYTECHNICA-CIVIL ENGINEERING	PERIOD POLYTECH-CIV	0553-6626
PROBABILISTIC ENGINEERING MECHANICS	PROBABILIST ENG MECH	0266-8920
PROCEEDINGS OF THE INSTITUTION OF CIVIL ENGINEERS-TRANSPORT	P I CIVIL ENG-TRANSP	0965-092X
PROCEEDINGS OF THE INSTITUTION OF CIVIL ENGINEERS-CIVIL ENGINEERING	P I CIVIL ENG-CIV EN	0965-089X
PROCEEDINGS OF THE INSTITUTION OF CIVIL ENGINEERS-ENGINEERING SUSTAINABILITY	P I CIVIL ENG-ENG SU	1478-4629
QUALITY AND RELIABILITY ENGINEERING INTERNATIONAL	QUAL RELIAB ENG INT	0748-8017
SCIENCE BULLETIN	SCI BULL	2095-9273
SCIENCE CHINA-TECHNOLOGICAL SCIENCES	SCI CHINA TECHNOL SC	1674-7321
SIMULATION MODELLING PRACTICE AND THEORY	SIMUL MODEL PRACT TH	1569-190X
土木工程（中文期刊）		
土木工程学报	—	1000-131X
力学学报	—	0459-1879
水利学报	—	0559-935
建筑结构学报	—	1000-6869

续表

JOURNAL TITLE	JOURNAL ABBREVIATION	ISSN
力学进展	—	1000-0992
岩土力学	—	000-7598
岩石力学与工程学报	—	1000-6915
煤炭学报	—	0253-9993
振动工程学报	—	1004-4523
振动与冲击	—	1000-3835
岩土工程学报	—	1000-4548
硅酸盐学报	—	0454-5648
岩石学报	—	1000-0569
工程力学	—	1000-4750
土木建筑与环境工程	—	1006-7329
地下空间与工程学报	—	1673-0836

土木工程（建筑材料）

JOURNAL TITLE	JOURNAL ABBREVIATION	ISSN
CEMENT & CONCRETE COMPOSITES	CEMENT CONCRETE COMP	0958-9465
CEMENT AND CONCRETE RESEARCH	CEMENT CONCRETE RES	0008-8846
COMPUTERS AND CONCRETE	COMPUT CONCRETE	1598-8198
CONSTRUCTION AND BUILDING MATERIALS	CONSTR BUILD MATER	0950-0618
JOURNAL OF ADVANCED CONCRETE TECHNOLOGY	J ADV CONCR TECHNOL	1346-8014
JOURNAL OF COMPOSITES FOR CONSTRUCTION	J COMPOS CONSTR	1090-0268
JOURNAL OF MATERIALS IN CIVIL ENGINEERING	J MATER CIVIL ENG	0899-1561
MAGAZINE OF CONCRETE RESEARCH	MAG CONCRETE RES	0024-9831
MECHANICS OF MATERIALS	MECH MATER	0167-6636
SMART MATERIALS AND STRUCTURES	SMART MATER STRUCT	0964-1726

土木工程学科（岩土工程）

JOURNAL TITLE	JOURNAL ABBREVIATION	ISSN
SOIL MECHANICS AND FOUNDATION ENGINEERING	SOIL MECH FOUND ENG	0038-0741
SOILS AND FOUNDATIONS	SOILS FOUND	0038-0806
ACI MATERIALS JOURNAL	ACI MATER J	0889-325X
ENGINEERING GEOLOGY	ENG GEOL	0013-7952
ACTA GEOTECNICAL	ACTA GEOTECH	1861-1125
ADVANCES IN CONCRETE CONSTRUCTION	ADV CONCR CONSTR	2287-5301
BULLETIN OF ENGINEERING GEOLOGY AND THE ENVIRONMENT	B ENG GEOL ENVIRON	1435-9529

续表

JOURNAL TITLE	JOURNAL ABBREVIATION	ISSN
CANADIAN GEOTECHNICAL JOURNAL	CAN GEOTECH J	0008-3674
COMPUTERS & GEOSCIENCE	COMPUT GEOSCI-UK	0098-3004
COMPUTERS AND GEOTECHNICS	COMPUT GEOTECH	0266-352X
EUROPEAN JOURNAL OF SOIL SCIENCE	EUR J SOIL SCI	1351-0754
GEOMECHANICS AND ENGINEERING	GEOMECH ENG	2005-307X
GEOPHYSICAL RESEARCH LETTERS	GEOPHYS RES LETT	0094-8276
GEOSYNTHETICS INTERNATIONAL	GEOSYNTH INT	1072-6349
GEOTECHNICAL TESTING JOURNAL	GEOTECH TEST J	0149-6115
GEOTECHNIQUE	GEOTECHNIQUE	0016-8505
GEOTEXTILES AND GEOMEMBRANES	GEOTEXT GEOMEMBRANES	0266-1144
IEEE GEOSCIENCE AND REMOTE SENSING LETTERS	IEEE GEOSCI REMOTE S	1545-598X
INTERNATIONAL JOURNAL FOR NUMERICAL AND ANALYTICAL METHODS IN GEOMECHANICS	INT J NUMER ANAL MET	0363-9061
INTERNATIONAL JOURNAL OF GEOMECHANICS	INT J GEOMECH	1532-3641
INTERNATIONAL JOURNAL OF ROCK MECHANICS AND MINING SCIENCES	INT J ROCK MECH MIN	1365-1609
JOURNAL OF ASIAN EARTH SCIENCES	J ASIAN EARTH SCI	1367-9120
JOURNAL OF ROCK MECHANICS AND GEOTECHNICAL ENGINEERING	J ROCK MECH GEOTECH ENG	1674-7755
JOURNAL OF GEOTECHNICAL AND GEOENVIRONMENTAL ENGINEERING	J GEOTECH GEOENVIRON	1090-0241
PROCEEDINGS OF THE INSTITUTION OF CIVIL ENGINEERS-GEOTECHNICAL ENGINEERING	P I CIVIL ENG-GEOTEC	1353-2618
GEODERMA	GEODERMA	0016-7061
GEOLOGICAL SOCIETY OF AMERICA BULLETIN	GEOL SOC AM BULL	0016-7606
土木工程学科(结构工程)		
WIND AND STRUCTURES	WIND STRUCT	1226-6116
THIN-WALLED STRUCTURES	THIN WALL STRUCT	0263-8231
STRUCTURAL HEALTH MONITORING-AN INTERNATIONAL JOURNAL	STRUCT HEALTH MONIT	1475-9217
STRUCTURAL SAFETY	STRUCT SAF	0167-4730
STRUCTURE AND INFRASTRUCTURE ENGINEERING	STRUCT INFRASTRUCT E	1573-2479
STEEL AND COMPOSITE STRUCTURES	STEEL COMPOS STRUCT	1229-9367

续表

JOURNAL TITLE	JOURNAL ABBREVIATION	ISSN
STRUCTURAL AND MULTIDISCIPLINARY OPTIMIZATION	STRUCT MULTIDISCIP O	1615-147X
STRUCTURAL CONCRETE	STRUCT CONCRETE	1464-4177
STRUCTURAL CONTROL & HEALTH MONITORING	STRUCT CONTROL HLTH	1545-225
STRUCTURAL DESIGN OF TALL AND SPECIAL BUILDINGS	STRUCT DES TALL SPEC	1541-7794
STRUCTURAL ENGINEERING AND MECHANICS	STRUCT ENG MECH	1225-4568
STRUCTURAL ENGINEERING INTERNATIONAL	STRUCT ENG INT	1016-8664
EARTHQUAKE ENGINEERING & STRUCTURAL DYNAMICS	EARTHQ ENG STRUCT D	0098-8847
ACI STRUCTURAL JOURNAL	ACI STRUCT J	0889-3241
ARCHIVES OF CIVIL AND MECHANICAL ENGINEERING	ARCH CIV MECH ENG	1644-9665
ADVANCED STEEL CONSTRUCTION	ADV STEEL CONSTR	1816-112X
ADVANCES IN STRUCTURAL ENGINEERING	ADV STRUCT ENG	1369-4332
COMPOSITE STRUCTURE	COMPOS STRUCT	0263-8223
COMPUTERS & STRUCTURES	COMPUT STRUCT	0045-7949
ENGINEERING JOURNAL-AMERICAN INSTITUTE OF STEEL CONSTRUCTION	ENG J AISC	0013-8029
ENGINEERING STRUCTURES	ENG STRUCT	0141-0296
FRONTIERS OF STRUCTURAL AND CIVIL ENGINEERING	FRONT STRUCT CIV ENG	2095-2430
INTERNATIONAL JOURNAL OF CONCRETE STRUCTURES AND MATERIALS	INT J CONCR STRUCT M	1976-0485
INTERNATIONAL JOURNAL OF SOLIDS AND STRUCTURES	INT J SOLIDS STRUCT	0020-7683
INTERNATIONAL JOURNAL OF STRUCTURAL STABILITY AND DYNAMICS	INT J STRUCT STAB DY	0219-4554
INTERNATIONAL JOURNAL OF STEEL STRUCTURES	INT J STEEL STRUCT	1598-2351
JOURNAL OF CONSTRUCTIONAL STEEL RESEARCH	J CONSTR STEEL RES	0143-974X
JOURNAL OF STRUCTURAL ENGINEERING	J STRUCT ENG	0733-9445
PROCEEDINGS OF THE INSTITUTION OF CIVIL ENGINEERS-STRUCTURES AND BUILDINGS	P I CIVIL ENG-STR B	0965-0911

土木工程学科(市政工程)

COMPUTERS & FLUIDS	COMPUT FLUIDS	0045-7930

JOURNAL TITLE	JOURNAL ABBREVIATION	ISSN
EXPERIMENTS IN FLUIDS	EXP FLUIDS	0723-4864
INTERNATIONAL JOURNAL FOR NUMERICAL METHODS IN FLUIDS	INT J NUMER METH FL	0271-2091
JOURNAL OF FLUID MECHANICS	J FLUID MECH	0022-1120
JOURNAL OF FLUIDS AND STRUCTURES	J FLUID STRUCT	0889-9746
PROCEEDINGS OF THE INSTITUTION OF CIVIL ENGINEERS-MUNICIPAL ENGINEER	P I CIVIL ENG-MUNIC	0965-0903
PROCEEDINGS OF THE INSTITUTION OF CIVIL ENGINEERS-WATER MANAGEMENT	P I CIVIL ENG-WAT M	1741-7589
QUARTERLY JOURNAL OF ENGINEERING GEOLOGY AND HYDROGEOLOGY	Q J ENG GEOL HYDROGE	1470-9236
ROCK MECHANICS AND ROCK ENGINEERING	ROCK MECH ROCK ENG	0723-2632
土木工程学科（桥梁与隧道工程）		
TUNNELLING AND UNDERGROUND SPACE TECHNOLOGY	TUNN UNDERGR SP TECH	0886-7798
JOURNAL OF BRIDGE ENGINEERING	J BRIDGE ENG	1084-0702
土木工程学科（防灾减灾工程与防护工程）		
BULLETIN OF EARTHQUAKE ENGINEERING	B EARTHQ ENG	1570-761X
BULLETIN OF THE SEISMOLOGICAL SOCIETY OF AMERICA	B SEISMOL SOC AM	0037-1106
EARTHQUAKE ENGINEERING AND ENGINEERING VIBRATION	EARTHQ ENG ENG VIB	1671-3664
EARTHQUAKE ENGINEERING AND STRUCTURAL DYNAMICS	EARTHQ ENG STRUCT D	0098-8847
EARTHQUAKE SPECTRA	EARTHQ SPECTRA	8755-2930
EARTHQUAKES AND STRUCTURES	EARTHQ STRUCT	2092-7614
INTERNATIONAL JOURNAL OF CRITICAL INFRASTRUCTURE PROTECTION	INT J CRIT INFR PROT	1874-5482
JOURNAL OF EARTHQUAKE AND TSUNAMI	J EARTHQ TSUNAMI	1793-4311
JOURNAL OF EARTHQUAKE ENGINEERING	J EARTHQ ENG	1363-2469
NATURAL HAZARDS	NAT HAZARDS	0921-030X
NATURAL HAZARDS AND EARTH SYSTEM SCIENCES	NAT HAZARD EARTH SYS	1561-8633
SEISMOLOGICAL RESEARCH LETTERS	SEISMOL RES LETT	0895-0695
SHOCK AND VIBRATION	SHOCK VIB	1070-9622
SOIL DYNAMICS AND EARTHQUAKE ENGINEERING	SOIL DYN EARTHQ ENG	0267-7261

续表

JOURNAL TITLE	JOURNAL ABBREVIATION	ISSN
土木工程学科（供热、供燃气、通风及空调工程）		
BUILDING AND ENVIRONMENT	BUILD ENVIRON	0360-1323
CIVIL ENGINEERING AND ENVIRONMENTAL SYSTEMS	CIV ENG ENVIRON SYST	1028-6608
ENERGY AND BUILDINGS	ENERG BUILDINGS	0378-7788
JOURNAL OF BUILDING PERFORMANCE SIMULATION	J BUILD PERFORM SIMU	1940-1493
PROCEEDINGS OF THE INSTITUTION OF CIVIL ENGINEERS-ENERGY	P I CIVIL ENG-ENERGY	1751-4223

第 15 章 论文检索与管理软件使用指南

第 8 章"文献检索与管理"中提到,科研人员可根据需求选择文献检索方式。其中,主题检索是最常用的检索方式,用于获取某个研究领域的文献。选择主题、标题等字段,输入检索词或检索式,其中重要的概念可以选择在标题字段检索。

我们以具体的科研课题为例,介绍论文检索与管理软件的使用。

15.1 课题介绍

我们选择一个土木工程领域的课题作为检索实例。课题信息如下:

(1) 课题名称:裂隙岩体多场耦合效应与调控。

(2) 课题简介:由于深部岩体的非线性力学特性和复杂的赋存环境,裂隙岩体多场耦合效应与调控是深地能源工程面临的关键难题。为揭示热水力化耦合作用下裂隙岩体力学特性劣化机制,拟建立裂隙变形/损伤-渗流-传热/传质耦合模型,发展基于现场示踪试验和等效渗流通道模型的参数反演方法以及多场耦合离散元数值模拟方法,提出井筒结构传热传质简化计算模型。研究成果将应用于京津冀地区地热资源评价与开采优化设计。

(3) 检索目的:查找裂隙岩体渗流传热传质的学术论文作为研究的参考文献。

15.2 论文检索步骤

1) 分析课题,提取主要概念

对于本课题,按照概念的重要程度及相互关系,主要分为以下几组:

(1) 研究对象:裂隙岩体。

(2) 参数/性质:渗流、传热、传质等力学特性。说明:力学特性是渗流、传热、传质的上位词。

(3) 研究方法/技术:离散元、剪切、应力效应、热-水-力-化耦合等。

(4) 其他:热储等。

2) 确定检索词

通过阅读相关文献、利用字词典、翻译助手等多种途径,选择、扩展中外文检索词。选择主要检索词如表 15.1 所示。

第15章 论文检索与管理软件使用指南

表 15.1 课题主要概念及其检索词

概念组序号	主要概念	中文检索词	英文检索词
1	裂隙岩体	岩体裂隙、裂隙岩体	rock fracture、fractured rock、single fracture、fracture networks
2	渗流	渗流、流体流动、流体流	fluid flow、water flow、seepage
2	传热	传热	heat transfer、heat transport
2	传质	传质、溶质传输、溶质运移	solute transport、mass transport
2	力学特性	力学特性	mechanical behavior、mechanical properties、mechanical characteristics
3	离散元	离散元、离散单元	discrete element model、DEM
3	剪切	剪切、直剪	shear、direct shear
3	应力效应	应力效应、应力状态效应	stress effect、stress state effects
4	热-水-力-化耦合	热-水-力-化耦合	thermal-hydrological-mechanical-chemical、thermal-hydro*-mechanical-chemical、thermal-hydraulic-mechanical-chemical、THMC
4	水-力-化耦合	水-力-化耦合、流固化学耦合	hydro-mechanical-chemical
4	多场耦合	多场耦合	multi-field coupling、multifield coupling
5	热储	热储、地热田、地热储、地热资源、地热系统	geothermal reservoir、geothermal wells、geothermal storing

3) 检索式的确定

根据检索词之间的逻辑关系,编制检索式。其中:不同组概念之间用"AND(与)"连接;同组概念的同义词、近义词、缩写等用"OR(或)"连接;用逻辑运算符"NOT(非)"排除某个概念(表15.2)。

表 15.2 检索式及其检索结果

序号	检索式	检索结果(20200610 检索)			
		WOS	WOSCC	SCIE+CPCI-S	SCIE
S1	TS=("rock*fracture*" OR "fractured rock*" OR "single fracture" OR "fracture* network*") 中文释义:主题=裂隙岩体	13997	9671	9174	7962
S2	TI=("rock*fracture*" OR "fractured rock*" OR "single fracture" OR "fracture* network*") 中文释义:标题=裂隙岩体	4112	2815	2656	2254

续表

序号	检索式	检索结果（20200610 检索）			
		WOS	WOSCC	SCIE+CPCI-S	SCIE
S3	TS=("fluid flow" OR "water flow" OR seepage OR "heat transfer" OR "heat transport" OR "solute transport" OR " mass transport " OR " mechanical behavior * " OR "mechanical propert * " OR "mechanical characteristics") 中文释义：主题=(渗流 OR 传热 OR 传质 OR 力学特性)	1 821 781	798 038	770 218	665 465
S4	TS=("discrete element" OR DEM OR shear OR "stress * effect * " OR " stress * state effect * " OR thermal-hydro * -mechanical-chemical OR THMC OR thermal-hydraulic-mechanical-chemical OR hydro * -mechanical-chemical) 检索式中文释义：主题=(离散元 OR 剪切 OR 应力效应 OR 热水力化)	908 662	468 077	453 252	398 972
S5=S1 AND S3	TS=(" rock * fracture * " OR "fractured rock * " OR "single fracture" OR "fracture * network * ") AND TS=("fluid flow" OR " water flow" OR seepage OR "heat transfer" OR "heat transport" OR "solute transport" OR " mass transport " OR " mechanical behavior * " OR "mechanical propert * " OR "mechanical characteristics") 检索式中文释义：主题=裂隙岩体 AND 主题=(渗流 OR 传热 OR 传质 OR 力学特性)	4345	3276	3117	2776
S6=S2 AND S3	TI=(" rock * fracture * " OR "fractured rock * " OR "single fracture" OR "fracture * network * ") AND TS=("fluid flow" OR "water flow" OR seepage OR "heat transfer" OR "heat transport" OR "solute transport" OR " mass transport " OR " mechanical behavior * " OR "mechanical propert * " OR "mechanical characteristics") 检索式中文释义：标题=裂隙岩体 AND 主题=(渗流 OR 传热 OR 传质 OR 力学特性)	1495	1066	994	853
S7=S1 AND S3 AND S4	TS=(" rock * fracture * " OR "fractured rock * " OR "single fracture" OR "fracture * network * ") AND TS=("fluid flow" OR "water flow" OR seepage OR "heat transfer" OR "heat transport" OR "solute transport" OR " mass transport " OR " mechanical behavior * " OR "mechanical propert * " OR "mechanical characteristics") AND TS=("discrete element" OR DEM OR shear OR "stress * effect * " OR " stress * state effect * " OR thermal-hydro * -mechanical-chemical OR THMC OR thermal-hydraulic-mechanical-chemical OR hydro * -mechanical-chemical) 检索式中文释义：主题=裂隙岩体 AND 主题=(渗流 OR 传热 OR 传质 OR 力学特性) AND 主题=(离散元 OR 剪切 OR 应力效应 OR 热水力化)	715	553	522	469

续表

序号	检索式	检索结果（20200610 检索）			
		WOS	WOSCC	SCIE+CPCI-S	SCIE
S8=S2 AND S3 AND S4	TI=("rock * fracture * " OR "fractured rock * " OR "single fracture" OR "fracture * network * ") AND TS=("fluid flow" OR "water flow" OR seepage OR "heat transfer" OR "heat transport" OR "solute transport" OR "mass transport" OR "mechanical behavior * " OR "mechanical propert * " OR "mechanical characteristics") AND TS=("discrete element" OR DEM OR shear OR "stress * effect * " OR "stress * state effect * " OR thermal-hydro * -mechanical-chemical OR THMC OR thermal-hydraulic-mechanical-chemical OR hydro * -mechanical-chemical)	259	204	183	156
	检索式中文释义：标题＝裂隙岩体 AND 主题＝（渗流 OR 传热 OR 传质 OR 力学特性）AND 主题＝（离散元 OR 剪切 OR 应力效应 OR 热水力化）				

检索说明：

（1）字段标识。TS 为主题字段，TI 为标题字段。WOS 平台中，主题字段包括标题、摘要、关键词等字段。同一检索式，选择不同字段，检索结果不同。在主题字段中检索，检索到的文献数量多于标题字段。如在主题字段中检索的 S1，比在标题字段中检索的 S2 数量要多。

（2）数据库。不同的数据库，检索到的文献有差异。对于本课题，除了 Web of Science 平台，还可以检索 Engineering Village 平台、Scopus 等资源。上述检索例子中，WOS 为 Web of Science 平台所有数据库；WOSCC 为 Web of Science Core Collection（Web of Science 核心合集）；SCIE＋CPCI-S 是指被 Web of Science 核心合集中的两个子库——Science Citation Index Expanded（SCI-EXPANDED，科学引文索引）收录的期刊论文及被 Conference Proceedings Citation Index-Science（CPCI-S）收录的会议论文；SCIE 是指被科学引文索引收录的期刊论文。

WOS 平台数据库及检索到的文献关系如下：Web of Science 平台所有数据库→Web of Science 核心合集→SCIE＋CPCI-S→SCIE（图 15.1）。

检索结果说明：

由表 15.2 可以看出：

（1）将第一重要概念"裂隙岩体"放在标题字段检索（S2：TI＝("rock * fracture * " OR "fractured rock * " OR "single fracture" OR "fracture * network * ")），检索得到文献的查准率比在主题字段（S1）中检索更高。S2 检索到的是关于裂隙岩体各方面研究的文献。如果要全面了解裂隙岩体的国内外研究概况，可以对 S2 检索到的几千篇文献进行系统分析。

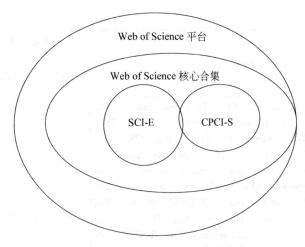

图 15.1　Web of Science 平台数据库关系

(2) 利用检索式 S6：TI=("rock * fracture * " OR "fractured rock * " OR "single fracture" OR "fracture * network * ") AND TS=("fluid flow" OR "water flow" OR seepage OR "heat transfer" OR "heat transport" OR "solute transport" OR "mass transport" OR "mechanical behavior * " OR "mechanical propert * " OR "mechanical characteristics")，检索到的是关于裂隙岩体渗流或传热传质等力学性质研究的文献。对这些论文进行分析，可以全面了解裂隙岩体渗流传热传质等方面的国内外研究概况。

(3) 利用检索式 S8：TI=("rock * fracture * " OR "fractured rock * " OR "single fracture" OR "fracture * network * ") AND TS=("Fluid flow" OR "water flow" OR Seepage OR "Heat transfer" OR "heat transport" OR "Solute transport" OR "mass transport" OR "Mechanical behavior * " OR "Mechanical Propert * " OR "Mechanical characteristics") AND TS=("Discrete element" OR DEM OR Shear OR "stress * effect * " OR " stress * state effect * " OR thermal-hydro * -mechanical-chemical OR THMC OR thermal-hydraulic-mechanical-chemical OR hydro * -mechanical-chemical)，检索到的是采用离散元或直剪、热水力化耦合等方法研究裂隙岩体渗流传热传质等力学特性的文献。用 S8 检索到的文献与本课题研究更为密切相关，对本研究有重要参考价值。我们以 S8 为例进行检索和文献分析。

15.3　论文检索结果及分析

1) 论文检索

采用检索式 S8，选择"SCIE＋CPCI-S"子库，选择高级检索（图 15.2）或基本检索（图 15.3）模式，检索得到 183 篇论文。其中：SCIE 期刊论文 156 篇，会议论文 34 篇，有 7 篇论文同时被 SCIE 和 CPCI-S 收录。

图 15.2　高级检索模式

图 15.3　基本检索模式

2) 检索结果分析及应用

图 15.4 展示了文献检索结果，对检索结果可以进行排序、精炼、分析和输出、创建跟踪，

也可以查看具体文献的详细记录。

图 15.4　检索结果及结果处理

对检索到的论文进行分析,可以全面了解国内外研究概况,为学术研究提供参考。宏观分析:

(1) 技术发展趋势。

通过出版年分析,可以了解该领域的技术发展趋势和生命周期。通过分析,发现2011年以来,本研究领域处于稳步上升期,见图15.5。

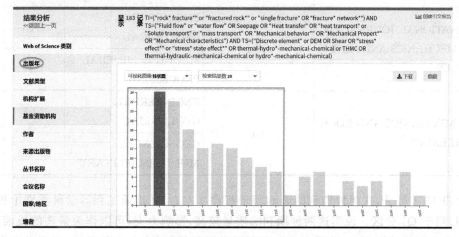

图15.5　检索结果分析(出版年)

(2) 主要国家/地区和研究机构。

按照国家/地区分析,可以发现中国、日本、美国、瑞典等国家在本技术领域研究较多。

按照机构分析,发现主要的研究机构有:China University of Mining Technology(中国矿业大学,20篇)、Nagasaki University(长崎大学,19篇)、Royal Institute of Technology(瑞典皇家理工学院,19篇)、Shandong University of Science Technology(山东科技大学,14篇)、United States Department of Energy Doe(美国能源部,13篇)、Lawrence Berkeley National Laboratory(劳伦斯伯克利国家实验室,11篇)、Wuhan University(武汉大学,11篇)、Kyoto University(京都大学,10篇)、Tsinghua University(清华大学,8篇)、Shaoxing University(绍兴文理学院,8篇)等。这些都是本学科领域要重点关注的研究机构。

(3) 主要科研人员。

按照作者分析,可以发现:绍兴文理学院土木工程学院李博教授(16篇)、中国矿业大学力学与土木工程学院刘日成研究员(14篇)、清华大学土木工程学院赵志宏副教授(12篇)、日本长崎大学蒋宇静教授(11篇)等学者是本学科领域的重要研究人员。未来的学术研究中可多关注这些领军人物的最新科研成果。

(4) 主要期刊。

通过对来源出版物进行分析,发现该领域的论文主要发表在 *International Journal of Rock Mechanics and Mining Sciences* (27篇)、*Rock Mechanics and Rock Engineering* (14篇)、*Journal of Geophysical Research Solid Earth* (8篇)、*Computers and Geotechnics* (6篇)、*Geofluids* (5篇)、*International Journal for Numerical and Analytical Methods in Geomechanics* (5篇)、*International Journal of Geomechanics* (5篇)等期刊上。其中 *International Journal of Rock Mechanics and Mining Sciences* 和 *Rock Mechanics and*

Rock Engineering 是发文最多的两种期刊,可以作为重要期刊进行论文跟踪,还可以作为备选投稿期刊。

通过 JCR 数据库,获取这两个期刊的影响因子、影响力排名、Q 值等信息(表 15.3)。

表 15.3 期刊 JCR 数据

期刊名称	影响因子(2018 年)	JCR 类别	排名(Rank)	分区
INTERNATIONAL JOURNAL OF ROCK MECHANICS AND MINING SCIENCES	3.769	ENGINEERING, GEOLOGICAL	5/38	Q1
		MINING & MINERAL PROCESSING	1/19	Q1
ROCK MECHANICS AND ROCK ENGINEERING	4.1	ENGINEERING, GEOLOGICAL	2/38	Q1
		GEOSCIENCES, MULTIDISCIPLINARY	23/196	Q1

由表 15.3 可见,这两种期刊的影响因子都在 4 左右,在地质工程等学科类别中排名位居前列,属于 Q1 分区。说明这两种期刊的质量较高、影响力大,可以作为备选的投稿期刊。但要注意选择投稿期刊时,除了考虑期刊所属学科领域、影响力等因素外,还要考虑审稿周期、录用率等诸多因素,应综合考虑。

15.4 文献管理

检索得到相关文献后,除了宏观分析外,还要深入挖掘论文内容,对文献进行管理,用于科研和论文写作等。

1)论文下载、阅读

选择感兴趣的论文,从全文数据库或免费资源下载原文,阅读分析。

2)导入文献管理软件

挑选需要的相关文献,选择打印、发送到邮箱、添加到标记结果列表、导入到个人文献管理软件等多种输出方式,对其进行综合管理和利用。

NoteExpress 或 EndNote Desktop(客户端版)是常用的文献管理软件。如果参考文献大多为外文文献,未来要发表外文学术论文,可以选择 EndNote。我们以 EndNote 客户端版为例:

(1)下载安装文献管理软件 EndNote 客户端版。

(2)创建个人文献数据库:File→New→创建新建数据库 Library(裂隙岩体.enl)。见图 15.6。

(3)获取书目数据

获取或添加文献记录,一般有以下几种方式:利用过滤器导入从数据库中检索得到的文献;导入已有的 PDF 全文;在线检索获取书目数据;手工输入文献题录信息。

以上述数据库检索为例,选择感兴趣的文献直接导入 EndNote 客户端,或者添加到标

第 15 章　论文检索与管理软件使用指南

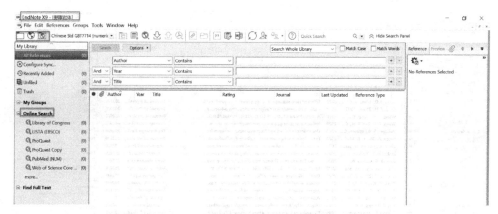

图 15.6　创建 EndNote 个人数据库

记结果列表中再选择输出方式导入 EndNote(图 15.7、图 15.8)。

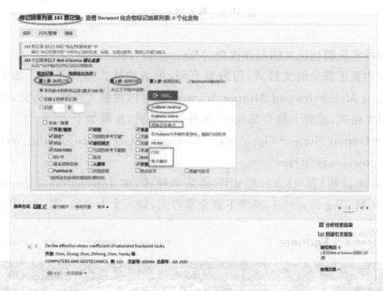

图 15.7　WOS 文献输出页面

3) 管理个人文献数据库

对 EndNote 数据库中的已有文献记录进行管理,如编辑、移动、删除、复制记录,为记录添加笔记、附件(全文、图片等),对记录进行查重、检索等。

选择 EndNote 中的某条记录,可以单击"Find Full Text"按钮在线下载全文(前提条件是具有全文访问权限),添加笔记等附件。如果不能在线下载全文,也可以将个人保存的 PDF 文档作为附件添加。还可以对整个数据库进行检索,如用作者"Zhao, Z. H"检索得到 12 条记录。还可以创建 Groups,将某些记录保存在 Group 中。

4) 论文写作

论文写作中,出版社对论文及参考文献有一定的格式要求。利用个人文献管理软件,除可从数据库中查找需要的参考文献外,还可以生成特定格式的引文列表,并可以在创作文档中直接插入特定出版社要求样式的参考文献、删除参考文献、修改参考文献样式。EndNote

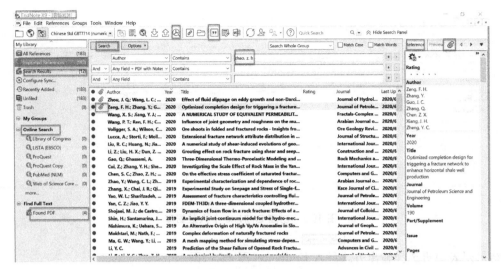

图 15.8 导入到 EndNote

还提供 300 多种常见期刊论文的写作模版(Manuscript Templates)。

EndNote 内置了部分引文样式,但数量有限,更多的样式需要下载。如 *International Journal of Rock Mechanics and Mining Sciences* 没有内置引文样式。为此,可以单独下载某个期刊的引文格式,或者下载全部 6000 多种引文格式,步骤如下:

① Edit→Output Style→Open Style Maneger→Get More on the Web→https://endnote.com/downloads/styles/。

② 按照页面说明(图 15.9),检索下载某个样式,如 *International Journal of Rock Mechanics and Mining Sciences*,或者下载全部样式,导入个人电脑(图 15.10)。

图 15.9 下载样式

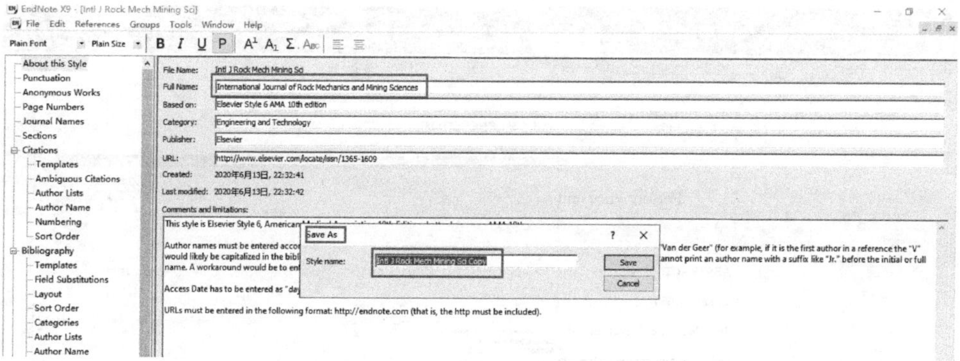

第 15 章 论文检索与管理软件使用指南

图 15.10 导入样式到 EndNote

5）写作引用

写作过程中，一般在 Introduction 中会对一些相关研究进行综述介绍，涉及的论著作为参考文献放在文后。利用文献管理软件可以在写作文档中直接插入出版社要求格式的引文。以在期刊 *International Journal of Rock Mechanics and Mining Sciences* 投稿为例，并在文中引用赵志宏老师的论文 *A mechanical-hydraulic-solute transport model for rough-walled rock fractures subjected to shear under constant normal stiffness conditions*。

期刊论文投稿说明一般可以在出版社网站查找，*International Journal of Rock Mechanics and Mining Sciences* 为爱思维尔出版的期刊，进入爱思维尔网站可以查看该期刊的相关信息（图 15.11）。

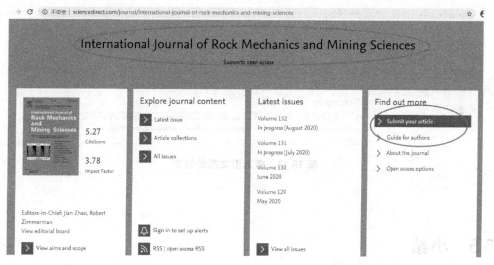

图 15.11 International Journal of Rock Mechanics and Mining Sciences 页面

论文引用：在学术论文 Word 文档中，单击 EndNote 工具栏，可以 Go To EndNote（进入 EndNote）；或者直接单击 Insert Citation，然后在 EndNote 数据库检索（图 15.12），单击 Insert 插入引文（图 15.13）。注意：如果 Word 文档中的默认引文格式不是我们要投稿期刊要求的格式，需要重新选择，如选择如图 15.10 所示的 Intl J Rock Mech Mining Sci 样式。

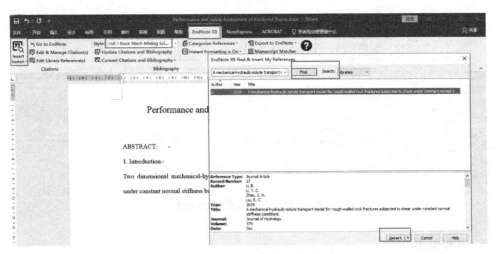

图 15.12　在 Word 文档中插入引文

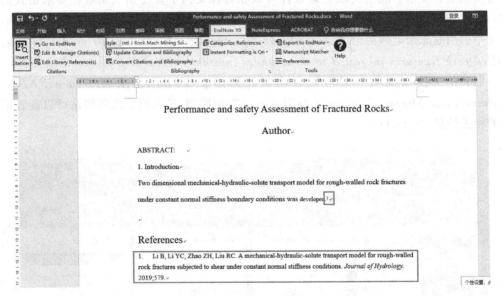

图 15.13　插入引文后的论文

15.5　小结

以具体课题为例,选择 Web of Science 平台和 EndNote,介绍了文献检索及其管理利用的大致流程。充分利用文献资源和个人文献管理工具,将有助于学术研究的开展,提高学术论文的写作水平,促进科研成果的交流和学科技术的发展。

第 16 章 学术共同体行为准则

学术论文的发表过程由作者、主编、审稿人、出版商共同合作完成,以下给出各机构或个人在论文撰写、投稿、审稿、出版等过程中应遵守的行为准则(翻译自 Elsevier Publishing Ethics,https://www.elsevier.com/about/policies/publishing-ethics)。图 16.1 是爱思维尔网站出版伦理。

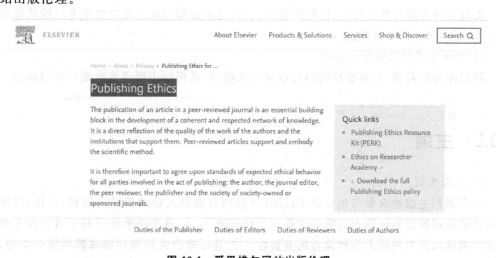

图 16.1 爱思维尔网站出版伦理

要建立连贯完备并能被广泛称赞的知识网络,发表文章、接受同行评审是非常必要的,这直接反映了作者及其赞助机构的工作质量。同行评审的文章为科学方法提供了支撑,具有使表达形象化的作用。故对出版过程的各职责方,包括作者、期刊主编、审稿人(同行评审)、出版商以及社会所有或私人赞助的期刊,制定统一的伦理道德准则非常重要。本章重点讨论各机构或个人在论文撰写、投稿、审稿、出版等过程中应遵守的行为准则。

16.1 出版商(以下为爱思维尔的出版商申明)

1) 学术记录的保护

出版商的重要作用是支持期刊主编的工作,以及审稿人员所承担的匿名志愿工作,以维护学术记录的完整性。道德准则主要是针对偶有发生的学术不端行为,学术道德审查体制现已非常成熟,这有利于规范学术活动。在学术交流中,出版商则扮演辅助角色,但务必要确保出版活动遵循学术规范[74-75]。

爱思维尔作为世界主导的期刊出版商之一,非常重视对学术记录的监护。期刊应被视为"科学纪要"的记录者,出版商就是"科学纪要"的监护人[76],应时刻恪守学术道德准则。

爱思维尔正在贯彻这些政策程序,以支持主编、审稿人和作者在这些方针下履行其道德义务,我们致力于与其他出版商和行业协会合作,为伦理道德问题、误差与勘误方面的最佳学术实践设定标准。

2) 自主编辑的维护

我们致力于确保潜在的广告、再版或其他商业收益不会对编辑决议产生影响。

3) 共同制定业界最佳学术实践

我们通过向主编提供出版伦理委员会(Committee on Publication Ethics)成员资格及所有提交给我们编辑系统的稿件的相似性对照检查报告,来推广最佳学术实践。

4) 为主编提供技术、程序及法律支持

我们支持主编与其他期刊、出版商进行交流,这个过程可在必要时提供专门的法律审查和法律顾问,对主编大有裨益。

5) 对研究者的出版准则培训

我们还为研究者,尤其是早期的行业研究人员,广泛提供出版道德准则培训与通告[77]。

16.2 主编

1) 出版决议

学术期刊主编通常要与相关学术社团(社会所有或私人资助赞助的期刊)合作,以便全权自主地决定提交给该期刊的哪些文章应予以发表[75,78],这类决策必须基于对正在考察的作品的确认以及对科研人员和读者的重要性。主编可能会受到期刊编辑部政策的指导,并受到针对诽谤、侵权和剽窃等问题有效的法律要求的约束。在做出这些决定时,主编可与其他主编或审稿人(或社团工作人员)协商。

2) 同行评审

主编应确保同行评审过程的公正、公正和及时,研究论文通常必须由至少两个外部审稿人独立审阅,必要时主编还应征询其他意见。

主编应选择相关领域中专业知识匹配的审稿人,并恪守学术活动规范,以避免选择不正当的同行审稿人[79]。主编应审查所有潜在的利益冲突,以及审稿人提出的自引建议,以确定是否存在潜在的不公正现象。

3) 公平竞争

主编只需评估稿件的知识内容,而不必考虑作者的种族、性别、性取向、宗教信仰、公民身份等。

期刊社论相关政策鼓励透明性、真实性、完整性,主编应确保审稿人和作者有明确的学术规范责任意识,期刊通信应基于标准电子提交系统。

主编应与出版商共同建立透明的机制,以便对审稿决议进行审查诉讼。

4) 刊物指标

主编不得通过人为增加期刊指标来影响期刊排名,特别是主编不得强行要求作者引用某一指定期刊的文章，并且不应要求作者引用主编的文章、成果和服务项目，以便从中牟利，除非出于真正的学术原因。

5）保密性

主编必须保护交予本期刊的所有材料，以及与审稿人通信的机密性，除非与相关作者和审稿人另有协议。在非常规情况下或须与出版商进行必要协商时，如认为有必要调查涉嫌学术不端的行为，主编可能会与其他期刊的主编共享有限的信息[80]。

主编必须保护审稿人的身份信息，除非是公开审稿和（或）审稿人同意披露其姓名。

未经作者明确书面同意，主编不得在自己的研究中揭露提交手稿中任何未发表的材料，通过同行评审获得的特权信息或想法必须保密，不得用于个人利益。

6）竞争利益声明

任何潜在的与主编相关的利益冲突，都应在主编任命前以书面形式向出版商声明，并在新的冲突出现时更新，出版商可在期刊中发布此类声明。

主编不得参与下述几类论文的通过决议：

他/她亲自撰写的论文；

其家庭成员、同事撰写的论文；

与主编可从中获益的成果或项目服务有关的论文。

此外，任何此类投稿都必须遵守该期刊的所有常规程序，同行评审必须独立于相关作者、编辑及其研究小组进行，并且要发表任何此类论文都必须对这些事项有明确声明[81]。

主编应采用爱思维尔有关作者和审稿人披露潜在利益冲突的政策，如 ICMJE① 准则[82]。

7）关注出版记录

主编应与出版商或学术社团共同审查、评估已报道或有嫌疑的学术不端行为（研究、出版物、审稿人和社论），以维护发表记录的完整性。

此类措施通常包括联系稿件或论文的作者，并适当考虑各自的投诉或主张，但也可能包括与相关机构和研究主体的进一步交流。主编应进一步合理利用出版商系统来监测学术不端行为，如剽窃。

收到确凿证据，指向学术不端时，主编应配合出版商［和（或）学术社团］及时安排发布对有关记录的更正、撤销或对其他更正和重要问题的解释说明[81]。

16.3 审稿人

1）对社论决议的贡献

同行评审有助于主编做出社论决议，并且通过与作者进行编辑交流也有助于作者改进论文。同行评议是正式学术交流的重要组成部分，是科学方法的核心。除规定伦理道德相关职责外，一般还要求审稿人推己及人对待其他作者及其作品，就像自己希望受到的待遇一样，遵守良好的评审规范。

任何选定的审稿人如果认为自己无资格评审稿件中报告的研究结果，或知道无法按时

---

① ICMJE：International Committee of Medical Journal Editors，国际医学期刊编辑委员会。

评审,应立即告知主编,并退出评审。

2) 保密性

收到的供评审的任何稿件必须视为机密文件,未经主编的许可,审稿人不得与任何人共同评审或共享有关论文的信息,或直接与作者联系。

一些主编鼓励与同事讨论或共同评审,但审稿人应先就此与主编讨论,以确保遵守保密规定,并确保参与者获得适当的信誉。

未经作者的明确书面同意,任何提交稿件中披露的未发表材料均不得在审稿人自己的研究中使用,通过同行评审获得的特权信息或想法必须保密,不得用于个人利益。

3) 警惕学术伦理问题

审稿人应警惕论文中潜在的伦理道德问题,并及时告知主编,如评审稿件与审稿人已知的任何其他已发表论文有大量实质性相似或重复,任何已经发表的观测、推导或论证的叙述均应附上相关引用。

4) 客观性准则与利益竞争准则

审稿过程应客观公正,审稿人应时刻留意他们可能产生的任何个人偏见,在审稿过程中引起重视,对作者的私人指责是不可取的。评判稿件要能明确表达审稿人的观点,并有充分的论据支持。

若审稿人因和与论文有关的任何作者、公司或机构间存在竞争、合作或其他关联而产生利益冲突,就必须在同意审阅该论文前,先征询主编的意见。

如果审稿人建议作者引用审稿人自己(或其同事)的作品,就必须出于真正的学术原因,而非为增加审稿人(或其同事)的引文计数,以提升其知名度。

16.4 作者

1) 报告标准

(1) 报告应真实准确列出所有作者并对应各自的工作内容,客观讨论各自工作的意义。

(2) 文章应准确展示原始数据。

(3) 文章应包含足够的研究细节和图文参考,以便读者能够重复该研究。包含虚假或故意含糊的陈述均构成学术不端,这是不可接受的。

(4) 综述和专业出版物文章的报告也应客观、准确,社论"意见"作品同样应明确认定。

2) 数据访问与保留

作者可能被要求提供支持其论文的研究数据,以供主编审查和(或)遵守期刊的数据公开要求。作者应在可能的情况下向公众提供这些数据,若条件允许,还应在论文发表后的合理年限内保留这些数据。关于详情作者可参考其期刊的《作者指南》。

3) 原创性与大众认可

作者应确保自己的作品完全原创,若作者使用了他人的作品和(或)文字,必须合理注明引证,必要时还须获得引用许可。

作者应引用对自己作品有真实影响的出版物,并对他人的工作给予适当的肯定,要在更大的学术记录中为引证作品提供适当的背景信息。未经信息来源明确书面许可,作者不得使用或报告通过私人方式获得的信息,如与第三方交流、通信或讨论。

学术成果剽窃形式多样,如顶替他人文章、复制或解释说明他人文章大量实质性内容、冒名宣布他人主导的研究成果。各种形式的学术剽窃均构成学术不端,是不可接受的。

4) 多次、赘余或并行出版

作者一般不能将本质上相同的研究,在一个以上主要出版物的期刊上发表。同时向多个期刊提交同一稿件构成学术不端,这是不可接受的。

作者通常不应将已发表论文提交另一种期刊评审,除非以下述形式:

(1) 摘要形式;

(2) 作已发表的演讲或学位论文的一部分;

(3) 电子预印本。

满足某些特定条件时,一稿多投是正当的,如:

(1) 文章类型为临床指南、翻译本等;

(2) 作者与期刊主编必须接收次级出版物,且次级出版物必须与原始文件共用相同的数据文件与原理解释;

(3) 次级出版物中必须引用原始参考文献。

一稿多投正当情形的详细情况可见 ICMJE 准则[82]。

5) 保密性

作品在保密服务期间透露出的信息不得擅自使用,除非有其作者明确的书面许可,例如:评审稿件过程、授权应用过程。

6) 论文作者

著作权应限于对报告所研究的概念提出、方法设计、过程操作或解释说明做出重大贡献的人,所有做出重大实质性贡献的人都应被列为合著者。

如果有人参与了论文的某些实质性方面的工作,如语言编辑、病历书写,则应在"致谢"部分予以认可。

通信作者应确保论文合著者的确定是正当的,并且所有合著者均已审阅并核准了文章最终版本,同意将其提交发表。

作者在提交稿件前,应仔细考虑作者名单和次序,并在初次投稿时提供确切的作者名单。只有在特殊情况下,主编才会在稿件提交后酌情考虑增加、删除或重新排列作者,且作者必须向主编明确注明此类请求,所有作者都必须同意此类添加、删除或重新安排。

作者对科研工作负有集体责任,每个作者都有责任保证与作品任何部分的准确性或完整性有关的问题能得到研究和妥善解决。

某些期刊可能具有特定的作者资格确定标准,如医学期刊可能遵循 ICMJE 作者身份定义[82]。作者应确保遵守相关期刊的政策。

7) 危害涉及人或动物主体

研究涉及任何异常危险的化学药品、程序或设备,作者都有义务在稿件中明确指出。

研究涉及人或动物主体的使用,作者都应确保在稿件中声明所有程序均遵守相关法律和制度方针,并已获得相应委员会批准。作者应在手稿中写明,已获得对人类受试者进行实验的知情同意,且必须始终遵守受试者的隐私权。

对于人类受试者,作者应确保所描述的工作遵循"世界医学协会"的《伦理道德准则》

(《赫尔辛基宣言》),该准则专门针对涉及人类的实验[83]。

所有动物实验均应遵循《ARRIVE 指南》和 1986 年版《英国动物(科学程序)法》和相关指南[84-85];或欧盟指令 2010/63/EU[86],该指令是关于保护出于科研探究而使用的动物;或美国有关人文关怀和实验动物使用准则的公共卫生服务政策。此外,如果适用,也可遵循《动物福利法》[87]。

如果作者希望在爱思维尔刊物中纳入案例详细信息或其他个人信息,如患者和任何其他个人的图像,就必须获得正当的公布管理许可。作者必须保留书面同意书,必要时需向爱思维尔提供同意书副本或已经获得许可的证据[88]。

8) 利益竞争宣告

世界医学编辑协会(World Association of Medical Editors, WAME)将利益冲突定义为,个人私利(竞争利益)与他(她)对学术、出版活动负有的职责间的背离,因此一个合格的遵守者可能会反思个人行为或评判是否出于他(她)对个人私利的考虑[89]。所有作者都应在稿件中公开与其他人或组织的任何可能对其工作产生负面影响,从而导致有失公允的财务和私人关系。

所有针对研究推进和(或)论文撰写的财务支持来源均应公开,如果有,还须公开赞助者在研究设计、收集、分析和解释说明原始数据、撰写报告以及制定论文提交决议等环节中的作用。如果资金来源未用于上述环节,应做出必要陈述。

应披露的潜在利益冲突的例子包括就业、顾问、股权、酬金、有偿专家证词、专利申请注册以及津贴或其他资金,潜在利益冲突应尽早披露[89]。

9) 基础错误通告

当发现自己发表的作品中存在重大误差或不精确性时,作者有义务立即通知期刊主编或出版商,并在主编认为必要时与主编合作撤回或更正论文。如果主编或出版商从第三方获悉已发表作品包含错误,则作者有义务与主编合作,如必要时向主编提供证据。

10) 图像完整性

在图像内增强效果、掩盖、移动、移除或引入特定功能是不可接受的,只要不掩盖或清除原稿中存在的任何信息,则可进行亮度、对比度或色彩平衡调整。尽管为提升图片清晰度而对图像进行处理是可以接受的,但出于其他目的对图片进行处理则可视为对科学伦理准则的滥用,应予以相应处理[90]。

作者应遵守相关期刊所用图形图像的特定政策,如提供原始图像作为论文补充材料,或将其存储在合适的存储库中。

11) 临床试验的透明度

爱思维尔支持临床试验透明,对相关期刊,作者应遵循临床试验注册和展示中的行业最佳标准,如 CONSORT① 指南,其在相关的期刊政策中有进一步陈述[82,91]。

---

① CONSORT: consolidated standards of reporting trials,临床试验报告统一标准。

附录 A　中英文对照表

中文名称	英文名称
引言	Introduction
方法	Methodology
结果	Results
讨论	Discussion
结论	Conclusion
标题	Title
作者信息	Author information
亮点	Highlights
摘要	Abstract
关键词	Keywords
图	Figures
表	Tables
致谢	Acknowledgement
参考文献	References
附录	Appendixes
补充材料	Supplementary materials
论文	Research article/Original Paper/Full paper
短文	Technical note/Short communication
案例研究	Case study
讨论论文	Letter/Closures/Errata
综述	Review article
会议论文	Conference paper
图形摘要	Graphic abstract
自述信	Cover letter
主编	Editor-in-chief
副主编	Associate Editor

续表

中文名称	英文名称
执行编辑	Managing Editors
总体概括	General comments
具体意见	Detailed/Technical comments
接收	Accept
小修	Minor revision
大修	Major revision
拒稿重投	Reject & resubmit
拒稿	Reject
审稿说明	A list of responses to the comments
注释版论文草稿	A separate copy of the revised paper in which you have marked the revisions made
大会	Symposium/Conference
专题研讨会	Workshop
大会报告	Plenary lecture
分组报告	Session presentation
口头报告	Oral presentation
海报	Poster
针对青年学者的大会报告专场	Plenary session for emerging scientists
幻灯片	Slides
书评	Book review
社论	Editorial/Editor's note
给编辑的信	Letter to Editor/Discussion and Closure
简历	Resume/Curricula vitae
开放获取	Open access
绿色开放获取	Green open access
黄金开放获取	Gold open access
文献	Document
中国知网	CNKI
中国科学引文数据库	CSCD
科学引文索引	Science Citation Index Expanded
美国土木工程师学会	American Society of Civil Engineers
英国土木工程师协会	Institution of Civil Engineers

续表

中文名称	英文名称
主题	Topic
地址	Address
机构扩展	Organizations-Enhanced
专利权人	ASSIGNEE NAME & CODE
发明人	Inventor
出版物名称	Publication Name
前期准备	Preparation
期刊	Journal
作者指导	Guide For Authors
投稿信	Cover letter
论文稿件	Manuscript
基金	Funding
利益冲突	Conflict of interest
稿件投递	Submission
注册	Register
邮箱	Email address
名	First name
姓	Family name
登录	Sign in
用户名	Username
密码	Password
文章类型	Regular issue/Special issue
职称	Title
通信作者	Corresponding author
主页	Home
编号	No.
状态	Current status
出版伦理委员会	Committee on Publication Ethics

附录 B CURRICULUM VITAE

(Updated on Aug. 10, 2020)

Personal Information

Name:	Zhihong ZHAO
Date of Birth:	March 22, 1983
Citizenship:	Chinese
Visiting Address:	He Shanheng Building 401
	Tsinghua University
	100084, Beijing China
Current Affiliation:	Department of Civil Engineering
	Tsinghua University, China
Telecommunication:	Tel.: +86 (10) 6279 5223
	E-mail: zhzhao@tsinghua.edu.cn

Education

Oct. 2008—Nov. 2011	PhD, Royal Institute of Technology (KTH), Sweden
	Thesis: Stress effects on solute transport in fractured rocks
	Advisors: Dr. Lanru Jing and Prof. Ivars Neretnieks
Sep. 2005—Jan. 2008	Master, University of Science and Technology Beijing, China
	Thesis: Improvement of limit equilibrium methods of slope analysis and engineering application
	Advisors: Prof. Jin-an Wang and Dr. Y.M. Cheng
Sep. 2001—Jul. 2005	Bachelor, University of Science and Technology Beijing, China
	Major in Civil Engineering

Work experience

Jul. 2017—	Associate professor at Department of Civil Engineering, Tsinghua University, China
Feb. 2014—Jun. 2017	Assistant professor at Department of Civil Engineering, Tsinghua University, China

Dec. 2011—Jun. 2014   Post-doc researcher at Department of Geological Sciences, Stockholm University, Sweden

Nov. 2006—Sep. 2008   Research assistant at Department of Civil and Structural Engineering, Hong Kong Polytechnic University

Research interests and projects
- Rock mechanics (Coupled THMC processes in fractured rocks)
- Underground engineering (EGS, Nuclear waste repositories, Deep tunnels)
- Numerical methods (Discrete element method, Particle mechanics method)

- Selected projects
1. Comprehensive evaluation technology of natural and artificial fracture networks in deep carbonate geothermal reservoirs, **National Key R&D Program of China**, PI, 2019—2022
2. Water-rock interaction induced multi-scale weakening of rock fractures and fracture shear strength criterion, **National Natural Science Foundation of China**, PI, 2018—2021
3. The mechanism and control of the abnormal deformation of super-high arch dams and mountains during the initial impoundment period, **National Natural Science Foundation of China**, co-PI, 2018—2022
4. The interaction mechanisms between dam and geological environment and control technologies for safety of high dam, **National Key R&D Program of China**, co-PI, 2018—2021
5. Discrete element modeling of thermal effects on flow and transport properties of fractured rocks in Enhanced Geothermal Systems (EGS), **National Natural Science Foundation of China**, PI, 2016—2018
6. Effects of reinjection on permeability of Beijing geothermal field: fundamental study and application, Beijing Natural Science Foundation, PI, 2015—2017
7. Stability and support technology of deep tunnels with large sections in high in-situ stress area, Tsinghua University Initiative Scientific Research Program, PI, 2015—2018
8. Numerical Simulation of geothermal fields for some key Sichuan-Tibet railway tunnels, Consulting Project, PI, 2019—2020
9. Numerical modeling on Xiaotangshan geothermal field in Beijing, Consulting Project, PI, 2019—2020
10. The optimized exploitation mode of geothermal resources in Tongzhou district of Beijing, Consulting Project, PI, 2019

Teaching and supervision
- Postgraduate course, Scientific writing and communication (2014—), Tsinghua University
- Undergraduate course, Introduction to underground space exploitation (2014—),

Tsinghua University
- Undergraduate course, Introduction to sponge city (2017—), Tsinghua University
- Tsinghua Postgraduate Education Reform Project, Scientific writing and communication, PI, (2017—2018)
- Postgraduate course, Applied geology, (2012, 2013), Stockholm University
- University pedagogy course (Passed, taken at Stockholm University and Tsinghua University)
- Postdoc and PhD students: Delei Shang (2018—), Sicong Chen (2018—), Haoran Xu (2019—), Yuedu Chen (2019)
- Master students, Huan Peng (2015—2018), Tiecheng Guo (2016—2019), Zihao Dou (2017—), Tao Lin (2019—)
- Bachelor students, Jiapeng Zhao (2016), Tiecheng Guo (2016), Sicong Chen (2018), Jingru Chen (2019), Tao Lin (2019)
- Visiting postgraduate students, Dong Zhou from University of Science and Technology Beijing (2015—2016), Guihong Liu from China University of Mining and Technology (2015—), Xue Li from China University of Geosciences Beijing (2016—2017), Haoran Xu from China University of Petroleum-Beijing (2017—2019), Xianxing Huang from China University of Petroleum-Beijing (2019—)

Awards and honors
- CouFrac2020 (International Conference on coupled processes in fractured geological media: observation, modeling, and application) Plenary Session of Emerging Scientists
- The 28th National Conference on Structural Engineering (China) Keynote lecture (2019)
- The 10th Asian Rock Mechanics Symposium (ARMS) Young Researcher Plenary Lecture (2018)
- KSRM Scholarship Award in 2017 Young Scholars' Symposium on Rock Mechanics (YSRM)
- Young Faculty Performance Award in Teaching, Tsinghua University (2016)
- The Young Elite Scientist Sponsorship (YESS) Program by China Association for Science and Technology (2015)
- The Recruitment Program of Beijing Youth Experts (2014)
- Chinese Government Award for Outstanding Self-financed Students Abroad (2011)

Publication
- *Books/Book chapters*

Zhao Z. Technical Regulation for Tracer Test in Geothermal Reservoirs, Head of drafting team, 2019 (in Chinese)

Zhao Z. Technical Regulation for Geothermal Water Reinjection in Sandstone Reservoirs, member of drafting team, DZ/T0330-2019 (in Chinese)

Zhao Z. 2017. Application of discrete element approach in fractured rock masses. In: Shao J. and Shojaei A (eds), Porous Rock Failure Mechanics: with Application to Hydraulic Fracturing, Drilling and Structural Engineering. Elsevier. pp 145-176.

- *Peer-reviewed journal papers (* corresponding author, # students)*

1. Li B., Ye X., Dou Z.#, **Zhao Z.***, Li Y., Yang Q. 2020. Shear strength of rock fractures under dry, surface wet and saturated conditions. Rock Mechanics and Rock Engineering. 53: 2605-2622.

2. **Zhao Z.***, Guo T.#, Li S., Wu W., Yang Q., Chen S#. 2020. Effects of joint surface roughness and orientational anisotropy on characteristics of excavation damage zone in jointed rocks. International Journal of Rock Mechanics and Mining Sciences. 128: 104265.

3. Peng H.#, **Zhao Z.***, Chen W., Chen Y., Fang Y., Li B. 2020. Thermal effect on permeability in a single granite fracture: Experiment and theoretical model. International Journal of Rock Mechanics and Mining Sciences. 131: 104358.

4. Liu Z., Xu H.#, **Zhao Z.***, Chen Z. 2019. DEM modeling of interaction between the propagating fracture and multiple pre-existing cemented discontinuities in shale. Rock Mechanics and Rock Engineering 52: 1993-2001.

5. Zhao X. G., Xu H. R.#, **Zhao Z.***, Guo Z., Cai M., Wang J. 2019. Thermal conductivity of thermally damaged Beishan granite under uniaxial compression. International Journal of Rock Mechanics and Mining Sciences. 115: 121-136.

6. Zuo J., Li Y., Zhang X., **Zhao Z.**, Wang T. 2018. The effects of thermal treatments on the subcritical crack growth of Pingdingshan sandstone at elevated high temperatures. Rock Mechanics and Rock Engineering. 51: 3439-3454.

7. **Zhao Z.***, Guo T.#, Ning Z., Dou Z., Dai F.*, Yang Q. 2018. Numerical modeling of stability of fractured reservoir bank slopes subjected to water-rock interactions. Rock Mechanics and Rock Engineering. 51: 2517-2531.

8. **Zhao Z.***, Peng H.#, Wu W., Chen Y-F. 2018. Characteristics of shear-induced asperity degradation of rock fractures and implications for solute retardation. International Journal of Rock Mechanics and Mining Sciences. 105: 53-61.

9. Zhao X., **Zhao Z.**, Guo Z., Cai M., Li X., Li P-F., Chen L., Wang J. 2018. Influence of thermal treatment on the thermal conductivity of Beishan granite. Rock Mechanics and Rock Engineering. 51: 2055-2074.

10. **Zhao Z.***, Liu Z., Pu H., Li X.# 2018. Effect of thermal treatment on brazilian tensile strength of granites with different grain size distributions. Rock Mechanics and Rock Engineering. 51: 1293-1303.

11. **Zhao Z.***, Yang J., Zhang D., Peng H.# 2017. Effects of wetting and cyclic wetting-drying on tensile strength of sandstone with a low clay mineral content. Rock

Mechanics and Rock Engineering. 50: 485-491.

12. Luo S., **Zhao Z.***, Peng H.#, Pu H. 2016. The role of fracture surface roughness in macroscopic fluid flow and heat transfer in fractured rocks. International Journal of Rock Mechanics and Mining Sciences. 87: 29-38.

13. **Zhao Z.*** 2016. Thermal influence on mechanical properties of granite: A microcracking perspective. Rock Mechanics and Rock Engineering. 49: 747-762.

14. **Zhao Z.***, Song E-x. 2015. Particle mechanics modeling of creep behavior of rockfill materials under dry and wet conditions. Computers and Geotechnics. 68: 137-146.

15. **Zhao Z.***, Li B., Jiang Y. 2014. Effects of fracture surface roughness on macroscopic fluid flow and solute transport in fracture networks. Rock Mechanics and Rock Engineering. 47: 2279-2286.

16. **Zhao Z.***, Liu L., Neretnieks I., Jing L. 2014. Solute transport in a single fracture: impacted by chemically mediated changes. International Journal of Rock Mechanics and Mining Sciences. 66: 69-75.

17. Bidgoli M. N., **Zhao Z.**, Jing L. 2013. Numerical evaluation of strength and deformability of fractured rocks. Journal of Rock Mechanics and Geotechnical Engineering. 5: 419-430.

18. **Zhao Z.*** 2013. Gouge particle evolution in a rock fracture undergoing shear: A microscopic DEM study. Rock Mechanics and Rock Engineering. 46: 1461-1479.

19. **Zhao Z.***, Rutqvist J., Leung C., Hokr M., Liu Q., Neretnieks I., Hoch A., Havlíček J., Wang Y., Wang Z., Wu Y., Zimmerman R. 2013. Impact of stress on solute transport in a fracture network: A comparison study. Journal of Rock Mechanics and Geotechnical Engineering. 5: 110-123.

20. **Zhao Z.**,* Jing L., Neretnieks I. 2012. Particle mechanics model for the effects of shear on solute retardation coefficient in rock fractures. International Journal of Rock Mechanics and Mining Sciences. 52: 92-102.

- *Refereed Conference papers or Abstracts*

1. **Zhao Z.**, Liu G. 2018. An integrated reservoir and well model for geothermal field. The International Conference on Coupled Processes in Fractured Geological Media: Observation, Modeling, and Application, Wuhan, China.

2. **Zhao Z.**, Zhao X. 2018. Thermally-induced microcracking in granites: insights from SEM observation and DEM modeling. The 10th Asian Rock Mechanics Symposium, Singapore.

3. **Zhao Z.**, Peng H. 2018. Characteristics of regular rock joint damage induced by shear. The 3rd International Conference on Damage Mechanics Shanghai, China.

4. **Zhao Z.**, Peng H. Luo S., Li B. 2018. Coupled TH processes in rough fractured reservoirs. The 2nd International Discrete Fracture Network Engineering Conference, Seattle, Washington, USA.

5. **Zhao Z.**, Peng H. 2017. Experimental study of effects of asperity degradation on the solute retardation coefficient in granite fractures. YSRM2017 & 5th NDRM, Jeju, Korea.
6. **Zhao Z.**, Zhou Z. Pu H. 2016. Shear behavior of heat-treated fractures in Beishan granite. ARMS9, Bali, Indonesia.
7. **Zhao Z.**, Li B. 2015. On the role of fracture surface roughness in fluid flow and solute transport through fractured rocks. In: ISRM2015, Montreal, Canada.
8. **Zhao Z.** 2014. Insights from DEM Modeling of Thermal Effects on Mechanical Properties of Crystalline Rocks. ARMS8, Sapporo.
9. **Zhao Z.** 2013. Discrete element simulation of fracture surface damage and gouge particle evolution in a rock fracture segment. SinoRock2013, Shanghai, pp385-390.
10. **Zhao Z.**, Jing L., Neretnieks I. 2010. Stress effects on nuclide transport in fractured rocks: a numerical study. In: Proceeding of European rock mechanics symposium (EUROCK2010), Lausanne, Switzerland, pp 783-786.

Professional service
- *Commissions*

  ISRM Commission on Coupled Thermal-Hydro-Mechanical-Chemical Processes in Fractured Rock, Commission member

  ISRM Commission on Radioactive Waste Disposal, Commission member

  Chinese Society for Rock Mechanics & Engineering, Commission on physical and numerical modeling, General secretary (Associate)

  Chinese Society for Rock Mechanics & Engineering, Commission on young scientists, Commission member

- *Journals*

  Computers and Geotechnics, Editorial Board Member

  Rock Mechanics and Rock Engineering, Editorial Board Member

  Geosystem Engineering, Editorial Board Member

  Journal of Rock Mechanics and Geotechnical Engineering, Scientific Editor

- *Conferences*
1. 2019 Beijing Association for Science and Technology (BAST) policy making Salon—Beijing-Tianjin-Hebei Geothermal Resources Sustainable Development and Low-Carbon City Construction, Chair
2. 2018 ARMS10—the 10th Asian Rock Mechanics Symposium, Session Chairman
3. 2018 Workshop for Young Scientists from Both Sides of the Taiwan Strait, Hong Kong and Macao, Chair
4. 2017 the 15th International Conference of the International Association for Computer

Methods and Advances in Geomechanics (15th IACMAG), Local Organizing Committee
5. 2017 Young Scientist Forum Supported by China Association for Science and Technology, Chair
6. 2016 ARMS9—the 9th Asian Rock Mechanics Symposium, Session Chairman
7. International Workshop on "Long-term Stability of High Dams" (Sinodam-2016), General Secretary

- *Reviewer for Technical Journals*:
  International Journal of Rock Mechanics & Mining Sciences, Rock Mechanics and Rock Engineering, Computers and Geotechnics, Hydrology and Earth System Sciences, Journal of Rock Mechanics and Geotechnical Engineering, Central European Journal of Geosciences, Engineering Computations, Canadian Geotechnical Journal, Geothermics, SpringerPlus, Hydrogeology Journal, Journal of Natural Gas Science & Engineering, Granular Matter, Applied Mathematical Modelling, Tectonophysics, Environmental Earth Sciences, Bulletin of Engineering Geology and the Environment, Applied Energy, Cold Regions Science and Technology

- *Others*
  Examiner of Shuang Luo's PhD thesis at Department of Civil Engineering, THU, 2018
  Anonymous examiner of PhD thesis at School of Civil Engineering, CSU, 2017
  Outstanding Student Paper Awards Judge of AGU Fall Meeting 2012
  Examiner of Barbara Kleine's Licenciate (Mid-term PhD) thesis at Department of Geological Sciences, SU (2012-12)
  Examiner of Madeleine Bohlin's Master thesis at Department of Geological Sciences, SU (2013-06)
  Examiner of Xiangyu Li's Master thesis at Department of Civil Engineering, THU (2015-06)
  European Research Council Advanced Grant 2016 Remote Review

Referees
1. Prof. Alasdair Skelton
   Department of Geological Sciences
   Director of the Bolin Centre for Climate Research
   Stockholm University
   Alasdair.Skelton@geo.su.se
   +46 (0)8 16 4750
2. Dr. Lanru Jing
   Division of Engineering Geology and Geophysics
   Department of Land and Water Resources Engineering

Royal Institute of Technology, Sweden
lanru@kth.se
+46 (0)8 790 6808

3. Prof. Ivars Neretnieks
Division of Chemical Engineering
Department of Chemical Engineering and Technology
Royal Institute of Technology, Sweden
niquel@kth.se
+46 (0)8 790 8229

参 考 文 献

[1] ZHAO Z H. Gouge particle evolution in a rock fracture undergoing shear: a microscopic DEM study [J]. Rock Mechanics and Rock Engineering, 2013, 46(6): 1461-1479.

[2] ZHAO Z H, GUO T C, NING Z Y, et al. Numerical modeling of stability of fractured reservoir bank slopes subjected to water—rock interactions[J]. Rock Mechanics and Rock Engineering, 2018, 51 (8): 2517-2531.

[3] ZHAO P, KÜHN D, OYE V, et al. Evidence for tensile faulting deduced from full waveform moment tensor inversion during the stimulation of the Basel enhanced geothermal system[J]. Geothermics, 2014, 52: 74-83.

[4] ZHAO Z H, YANG J, ZHOU D, et al. Experimental investigation on the wetting-induced weakening of sandstone joints[J]. Engineering Geology, 2017, 225: 61-67.

[5] LIU Z N, XU H R, ZHAO Z H, et al. DEM modeling of interaction between the propagating fracture and multiple pre-existing cemented discontinuities in shale[J]. Rock Mechanics and Rock Engineering, 2019, 52(6): 1993-2001.

[6] ZHAO Z H, RUTQVIST J, LEUNG C, et al. Impact of stress on solute transport in a fracture network: A comparison study[J]. Journal of Rock Mechanics and Geotechnical Engineering, 2013, 5 (2): 110-123.

[7] WAGNER L, KLUCKNER A, MONSBERGER C M, et al. Direct and distributed strain measurements inside a shotcrete lining: Concept and realisation[J]. Rock Mechanics and Rock Engineering, 2020, 53(2): 641-652.

[8] WIDESTRAND H, BYEGÅRD J, CVETKOVIC V, et al. Sorbing tracer experiments in a crystalline rock fracture at Äspö (Sweden): 1. Experimental setup and microscale characterization of retention properties[J]. Water Resources Research, 2007, 43(10): 1-17.

[9] CVETKOVIC V, CHENG H, WIDESTRAND H, et al. Sorbing tracer experiments in a crystalline rock fracture at Äspö (Sweden): 2. Transport model and effective parameter estimation[J]. Water Resources Research, 2007, 43(11): 1-16.

[10] CVETKOVIC V, CHENG H. Sorbing tracer experiments in a crystalline rock fracture at Äspö (Sweden): 3. Effect of microscale heterogeneity[J]. Water Resources Research, 2008, 44(12): 1-11.

[11] ZHAO Z H, SKELTON A. An assessment of the role of nonlinear reaction kinetics in parameterization of metamorphic fluid flow[J]. Journal of Geophysical Research: Solid Earth, 2014, 119(8): 6249-6262.

[12] SKELTON A, ANDRÉN M, KRISTMANNSDÓTTIR H, et al. Changes in groundwater chemistry before two consecutive earthquakes in Iceland[J]. Nature Geoscience, 2014, 7(10): 752-756.

[13] LORIA A F R, OLTRA J V C, LALOUI L. Equivalent pier analysis of full-scale pile groups subjected to mechanical and thermal loads[J]. Computers and Geotechnics, 2020, 120: 103410.

[14] ZHAO Z H, LIU L C, NERETNIEKS I, et al. Solute transport in a single fracture with time-dependent aperture due to chemically medicated changes[J]. International Journal of Rock Mechanics and Mining Sciences, 2014, 66: 69-75.

[15] CHEN J, LUO Y Q, VAN GROENIGEN K J, et al. A keystone microbial enzyme for nitrogen control of soil carbon storage[J]. Science Advances, 2018, 4(8): eaaq1689.

[16] ALIYU M D, CHEN H P. Optimum control parameters and long-term productivity of geothermal reservoirs using coupled thermo-hydraulic process modelling[J]. Renewable Energy, 2017, 112: 151-165.

[17] PYRAK-NOLTE L J, NOLTE D D. Approaching a universal scaling relationship between fracture stiffness and fluid flow[J]. Nature Communications, 2016, 7(1): 1-6.

[18] ZHAO Z H. Thermal influence on mechanical properties of granite: a microcracking perspective[J]. Rock Mechanics and Rock Engineering, 2016, 49(3): 747-762.

[19] ZHAO Q, TISATO N, KOVALEVA O, et al. Direct Observation of Faulting by Means of Rotary Shear Tests Under X-Ray Micro-Computed Tomography[J]. Journal of Geophysical Research: Solid Earth, 2018, 123(9): 7389-7403.

[20] DOU Z H, GAO T C, ZHAO Z H, et al. The role of water lubrication in critical state fault slip[J]. Engineering Geology, 2020, 271: 1-9.

[21] ZHAO Z H, JING L R, NERETNIEKS I, et al. A new numerical method of considering local longitudinal dispersion in single fractures[J]. International Journal for Numerical and Analytical Methods in Geomechanics, 2014, 38(1): 20-36.

[22] ZHAO Z H, JING L R, NERETNIEKS I, et al. Analytical solution of coupled stress-flow-transport processes in a single rock fracture[J]. Computers & Geosciences, 2011, 37(9): 1439-1449.

[23] ZHAO Z H, JING L R, NERETNIEKS I, et al. Numerical modeling of stress effects on solute transport in fractured rocks[J]. Computers & Geotechnics, 2011, 38(2): 113-126.

[24] DAY R A, GASTEL B. How to write and publish a scientific paper (Fifthedition)[M]. Santa Barbara: Greenwood, 1998.

[25] ZHAO X G, XU H R, ZHAO Z H, et al. Thermal conductivity of thermally damaged Beishan granite under uniaxial compression[J]. International Journal of Rock Mechanics and Mining Sciences, 2019, 115: 121-136.

[26] ZHAO Z H, SKELTON A. Simultaneous calculation of metamorphic fluid fluxes, reaction rates and fluid—rock interaction timescales using a novel inverse modeling framework[J]. Earth and Planetary Science Letters, 2013, 373: 217-227.

[27] WU W, ZHAO Z H, DUAN K. Unloading-induced instability of a simulated granular fault and implications for excavation-induced seismicity[J]. Tunnelling and Underground Space Technology, 2017, 63: 154-161.

[28] LI B Q, DA SILVA B G, et al. Laboratory hydraulic fracturing of granite: acoustic emission observations and interpretation[J]. Engineering Fracture Mechanics, 2019, 209: 200-220.

[29] LIU G H, WANG G L, ZHAO Z H, et al. A new well pattern of cluster-layout for deep geothermal reservoirs: Case study from the Dezhou geothermal field, China[J]. Renewable Energy, 2020: 484-499.

[30] WANG G L, LIU G H, ZHAO Z H, et al. A robust numerical method for modeling multiple wells in city-scale geothermal field based on simplified one-dimensional well model[J]. Renewable Energy, 2019, 139: 873-894.

[31] PARK C H, BOBET A. Crack coalescence in specimens with open and closed flaws: a comparison [J]. International Journal of Rock Mechanics and Mining Sciences, 2009, 46(5): 819-829.

[32] ZHOU D, ZHAO Z H, LI B et al. Permeablity evolution of grout infilled fractures subjected to

triaxial compression with low confining pressure[J]. Tunneling and Underground Space Technology incorporating Trenchless Technology Research, 2020, 104: 103539.

[33] JING L R. A review of techniques, advances and outstanding issues in numerical modelling for rock mechanics and rock engineering[J]. International Journal of Rock Mechanics and Mining Sciences, 2003, 40(3): 283-353.

[34] HSIEH P A. Fundamentals of Rock Mechanics (Book Review)[J]. Geofluids, 2009, 9: 251-252.

[35] BARIA G. Editorial[J]. Rock Mechanics and Rock Engineering, 2012, 45: 647.

[36] GLYNN P D. Letter to the Editor[J]. Computer & Geosciences, 2005, 31: 1305-1307.

[37] KIM W, JEONG J P. Retraction Note to: Non-Bernoulli-Compatibility Truss Model for RC Members Subjected to Combined Action of Flexure and Shear Part II: Its Practical Solution[J]. KSCE Journal of Civil Engineering, 2019, 23(6): 2812-2812.

[38] JOSEPH P F, LOGHIN A. Letter to the Editor of Engineering Fracture Mechanics[J]. Engineering Fracture Mechanics, 2020, 232: 107049.

[39] ZHAO T, LIU W, ZHANG L, et al. RETRACTED: Cluster analysis of risk factors from near-miss and accident reports in tunneling excavation[J]. Journal of construction engineering and management, 2018, 144(6): 04018040.

[40] WALSH S D C, LOMOV I N. Micromechanical modeling of thermal spallation in granitic rock[J]. International Journal of Heat and Mass Transfer, 2013, 65: 366-373.

[41] PETERIE S L, MILLER R D, INTFEN J W, et al. Earthquakes in Kansas induced by extremely far-field pressure diffusion[J]. Geophysical Research Letters, 2018, 45(3): 1395-1401.

[42] KERANEN K M, WEINGARTEN M, ABERS G A, et al. Sharp increase in central Oklahoma seismicity since 2008 induced by massive wastewater injection[J]. Science, 2014, 345(6195): 448-451.

[43] WILLEMS C J L, NICK H M. Towards optimisation of geothermal heat recovery: An example from the West Netherlands Basin[J]. Applied Energy, 2019, 247: 582-593.

[44] VISHAL V, PRADHAN S P, SINGH T N. Tensile strength of rock under elevated temperatures [J]. Geotechnical and Geological Engineering, 2011, 29(6): 1127.

[45] DUAN K, JI Y L, XU N W, et al. Excavation-induced fault instability: possible causes and implications for seismicity[J].Tunnelling and Underground Space Technology, 2019, 92: 103041.

[46] NORDLUND E, ZHANG P, DINEVA S, et al. Impact of fire on the stability of hard rock tunnels in Sweden[M].Stockholm: Stiftelsen bergteknisk forskning-Befo, 2015.

[47] VOGLER D, SETTGAST R R, ANNAVARAPU C, et al. Experiments and simulations of fully hydro-mechanically coupled response of rough fractures exposed to high-pressure fluid injection[J]. Journal of Geophysical Research: Solid Earth, 2018, 123(2): 1186-1200.

[48] SIRDESAI N N, SINGH T N, RANJITH P G, et al. Effect of varied durations of thermal treatment on the tensile strength of red sandstone[J]. Rock Mechanics and Rock Engineering, 2017, 50(1): 205-213.

[49] DIAO Y, ESPINOSA-MARZAL R M. The role of water in fault lubrication[J]. Nature Communications, 2018, 9(1): 1-10.

[50] ZOU L, HÅKANSSON U, CVETKOVIC V. Cement grout propagation in two-dimensional fracture networks: Impact of structure and hydraulic variability[J]. International Journal of Rock Mechanics and Mining Sciences, 2019, 115: 1-10.

[51] XIE L, MIN K B, SONG Y. Observations of hydraulic stimulations in seven enhanced geothermal

system projects[J]. Renewable Energy, 2015, 79: 56-65.

[52] INDRARATNA B, PREMADASA W, BROWN E T, et al. Shear strength of rock joints influenced by compacted infill[J]. International Journal of Rock Mechanics and Mining Sciences, 2014, 70: 296-307.

[53] IRVINE D J, SIMMONS C T, WERNER A D, et al. Heat and solute tracers: How do they compare in heterogeneous aquifers? [J]. Groundwater, 2015, 53(S1): 10-20.

[54] WALSH S D C, LOMOV I, ROBERTS J J. Geomechanical modeling for thermal spallation drilling [J]. Geothermal Resources Council Transactions, 2011, 35: 277-282.

[55] PASSELÈGUE F X, BRANTUT N, MITCHELL T M. Fault reactivation by fluid injection: Controls from stress state and injection rate[J]. Geophysical Research Letters, 2018, 45(23): 12,837-12,846.

[56] WU W, REECE J S, GENSTERBLUM Y, et al. Permeability evolution of slowly slipping faults in shale reservoirs[J]. Geophysical Research Letters, 2017, 44(22): 11,368-11,375.

[57] ZHAO Z H, PENG H, WU W, et al. Characteristics of shear-induced asperity degradation of rock fractures and implications for solute retardation[J]. International Journal of Rock Mechanics and Mining Sciences, 2018, 105: 53-61.

[58] DE LA BERNARDIE J, BOUR O, LE BORGNE T, et al. Thermal attenuation and lag time in fractured rock: theory and field measurements from joint heat and solute tracer tests[J]. Water Resources Research, 2018, 54(12): 10,053-10,075.

[59] MAKASIS N, NARSILIO G A, BIDARMAGHZ A. A machine learning approach to energy pile design[J]. Computers and Geotechnics, 2018, 97: 189-203.

[60] DENG H, MOLINS S, STEEFEL C, et al. A 2.5 D reactive transport model for fracture alteration simulation[J]. Environmental Science & Technology, 2016, 50(14): 7564-7571.

[61] GAN Q, ELSWORTH D. Production optimization in fractured geothermal reservoirs by coupled discrete fracture network modeling[J].Geothermics, 2016, 62: 131-142.

[62] GAN Q, ELSWORTH D. Analysis of fluid injection-induced fault reactivation and seismic slip in geothermal reservoirs[J]. Journal of Geophysical Research: Solid Earth, 2014, 119(4): 3340-3353.

[63] NASSERI M H B, SCHUBNEL A, YOUNG R P. Coupled evolutions of fracture toughness and elastic wave velocities at high crack density in thermally treated Westerly granite[J]. International Journal of Rock Mechanics and Mining Sciences, 2007, 44(4): 601-616.

[64] FANG Y, ELSWORTH D, WANG C, et al. Frictional stability-permeability relationships for fractures in shales[J]. Journal of Geophysical Research: Solid Earth, 2017, 122(3): 1760-1776.

[65] ZIJLSTRA W, HOF A L. Displacement of the pelvis during human walking: experimental data and model predictions[J]. Gait & Posture, 1997, 6(3): 249-262.

[66] PATEL S, MARTIN C D. Application of flattened Brazilian test to investigate rocks under confined extension[J]. Rock Mechanics and Rock Engineering, 2018, 51(12): 3719-3736.

[67] TATONE B S A, GRASSELLI G. An investigation of discontinuity roughness scale dependency using high-resolution surface measurements[J]. Rock Mechanics and Rock Engineering, 2013, 46 (4): 657-681.

[68] PYRAK-NOLTE L J, MORRIS J P. Single fractures under normal stress: The relation between fracture specific stiffness and fluid flow[J]. International Journal of Rock Mechanics and Mining Sciences, 2000, 37(1-2): 245-262.

[69] SO O Y, SCARAFIA L E, MAK A Y, et al. The dynamics of prostaglandin H synthases studies

with prostaglandin H synthase 2 Y355F unmask mechanisms of time-dependent inhibition and allosteric activation[J]. Journal of Biological Chemistry, 1998, 273(10): 5801-5807.

[70] GOEBEL T H W, BRODSKY E E. The spatial footprint of injection wells in a global compilation of induced earthquake sequences[J]. Science, 2018, 361(6405): 899-904.

[71] SOMERTON W H. Thermal properties and temperature-related behavior of rock/fluid systems[M]. New York: Elsevier, 1992.

[72] SOSNIN S, KARLOV D, TETKO I V, et al. Comparative study of multitask toxicity modeling on a broad chemical space[J]. Journal of Chemical Information and Modeling, 2018, 59(3): 1062-1072.

[73] DENG H, PETERS C A. Reactive transport simulation of fracture channelization and transmissivity evolution[J]. Environmental Engineering Science, 2019, 36(1): 90-101.

[74] The STM trade Association. International Ethical Principles for Scholarly Publication[EB/OL]. [2020-11-20]. https://www.stmassoc.org/2013_05_21_STM_E-thical_Principles_for_Scholarly_Publication.pdf.

[75] COPE. Codes of Conduct[EB/OL]. [2020-11-20]. https://publicationethics.org/re-sources/code-conduct.

[76] Elsevier. Elsevier policy on the permanence of the scientific record[EB/OL]. [2020-11-20]. https://www.elsevier.com/about/policies/article-withdrawal.

[77] Elsevier. Elsevier educational content on Ethics in Research & Publication[EB/OL]. [2020-11-20]. https://www.publishingcampus.elsevier.com/ethics.

[78] Elsevier. Elsevier policy on editorial independence[EB/OL]. [2020-11-20]. https://www.elsevier.com/about/poli-cies/editorial-independence

[79] World Medical Association (WMA). World Association of Medical Editors (WAME) Best Practice [EB/OL]. [2020-11-20]. http://www.wam-e.org/about/policy-statements#Best Practices for Peer Reviewer Selection

[80] Committee on Publication Ethics (COPE). Guidelines on Editors in Chief sharing[EB/OL]. [2020-11-20]. https://publicationethics.org/files/Sharing%20_of_In-formation_Among_EiCs_guidelines_web_version_0.pdf

[81] Elsevier. Elsevier's Publishing Ethics Resource Kit for Editors[EB/OL]. [2020-11-20]. https://www.else-vier.com/editors/perk

[82] International Committee of Medical Journal Editors. ICMJE Uniform requirements for manuscripts submitted to biomedical journals[EB/OL]. [2020-11-30]. http://www.icmje.org/

[83] World Medical Association (WMA). Helsinki Declaration for Medical Research in Human Subject [EB/OL]. [2020-11-20]. https://www.wma.net/policies-post/wmadec-laration-of-helsinki-ethical-principles-for-medical-research-involving-human-subjects

[84] Animal Research: Reporting of In Vivo Experiments (ARRIVE). ARRIVE Guidelines[EB/OL]. [2020-11-20]. https://www.nc3rs.org.uk/arrive-guidelines

[85] The U.K. Animals (Scientific Procedures) Act 1986[EB/OL]. [2020-11-20]. https://www.gov.uk/government/uploads/system/uploads/attachment_data/file/308593/ConsolidatedASPA1Jan2013.pdf

[86] European Commission. EU Directive 2010/63/EU for animal experiments[EB/OL]. [2020-11-20]. https://ec.europa.eu/en-vironment/chemicals/lab_animals/legislati-on_en.htm

[87] U.S. Public Health Service Policy on Humane Care and Use of Laboratory Animals[EB/OL]. [2020-11-20]. https://grants.nih.gov/grants/olaw/references/ph-spolicylabanimals.pdf

[88] Elsevier. Elsevier policy on patient consent[EB/OL]. [2020-11-20]. https://www.elsevier.com/

about/poli-cies/patient-consent

[89] World Medical Association (WMA). WAME Editorial statement on COI[EB/OL]. [2020-11-20]. http://www.wame.org/about/conflict-of-interest-in-peer-reviewed-medical

[90] ROSSNER M, YAMADA K M. What's in a picture? The temptation of image manipulation[J]. Journal of Cell Biology, 2004, 166(1): 11-15.

[91] CONSORT. CONSORT standards for randomized trials[EB/OL]. [2020-11-20]. http://www.consort-statement.org/